Information Series 33

Gold Panning and Placering in Colorado

How and Where

By Ben H. Parker, Jr.

Drawings and maps drafted by Larry Scott
Digital version recompiled by Larry Scott

Bill Ritter Jr., Governor
State of Colorado

Harris D. Sherman, Executive Director
Department of Natural Resources

Vincent Matthews
State Geologist and Director
Colorado Geological Survey

Colorado Geological Survey
Department of Natural Resources
Denver, Colorado
2009

CONTENTS

Beacon Hill Mine, Cripple Creek District

Idaho Springs District

Revenue Mine, Ouray District

Lewis Mine, Telluride District

Crystal Mill, Gunnison County

FIGURES

Red Mountain Pass barracks, Ouray County

TABLES

ABOUT THE AUTHOR

Although Dr. Ben H. Parker, Jr. was born in Oklahoma in 1928, most of his schooling occurred in Golden, Colorado where he later attended the Colorado School of Mines. He received his Geological Engineer degree in 1949 and his Doctor of Science (Geology) degree in 1961. His doctoral thesis "The Geology of the Gold Placers of Colorado", the most comprehensive document ever published on Colorado's gold placers, has guided gold explorationists and hobbyists for over two decades.

Ben's work in the geology of metallic and industrial mineral deposits and petroleum exploration extended throughout the United States, particularly in the Rocky Mountains and Alaska. His 18 years of overseas mineral exploration included extended periods in Mexico, Colombia, Brazil, Argentina, the Philippines, China, Greece and Turkey. His supervision led to discovery of the major Pachon porphyry copper deposit in Argentina.

Ben is a Registered Professional Engineer and Land Surveyor in Colorado and a Registered Geologist in California. He is a Certified Professional Geologist of the American Institute of Professional Geologists and held a commission as Mineral Surveyor of the U.S. Bureau of Land Management for over 10 years. Ben authored "Gold Placers of Colorado," published in 1974 in the Colorado School of Mines Quarterly and various short articles. Dr. Parker conducts a world-wide mineral exploration consulting practice from his home in Golden, Colorado.

FOREWORD

The Colorado Geological Survey is pleased to present a new, colorized version of *Gold Panning and Placering in Colorado–How and Where by Ben H. Parker*. The original and popular 1992 book has sold out prompting a new, digital version. The text is taken directly from the original version with just a few minor edits and updates. The maps were recreated using a digital elevation model (DEM) base to enhance detail and clarify topographic position.

Many placer deposits still have active claims on them. This means that any gold in the deposit is owned by the mineral claim owner. You must have permission of both the surface owner and the mineral owner in order to legally pan the deposit. The county courthouse is the official location for determining who owns the surface and who owns the mineral (gold) rights.

An excellent way to get involved in panning is to join a prospecting club. They usually own claims where members can legally pan. It is also useful to begin panning with experienced panners who can show you their tricks and share their experiences with you.

Explanation of Index Map Symbols

Location and page number of each index map is shown on state index map (pages xii-xiii).

36	Section of township and range identified in book
19	Unsurveyed section of township in range
	Town boundary or area of population
	Small towns or population place
	Abandoned or historic townsite
	Interstate or Federal highway
	State highway
	Other paved roads (local)
	Improved gravel road (access roads)
4WD	Unimproved gravel or dirt road (some are marked 4WD)
	County boundary
	Dry gulch or seasonal creek
	Stream or creek
	River
	Named formation or outcrop
	Named placer site
	Dredged tailings piles
70	Interstate highway
40	U. S. highway
119	State highway
129	County road
FR-406	U.S.F.S. route
	Significant mountain or peak
	Local hill or elevation point
	Lake or pond
Res	Reservoir

Explanation of Township and Range Survey System

The index maps located throughout this book are designed to assist the reader in locating all specific sites mentioned in the text. By utilizing the township and range surveying system which is used in the western half of the United States, all gold panning and placering sites can be found within one square-mile sections; and each section may be further subdivided into halves and quarters as shown in the diagrams below:

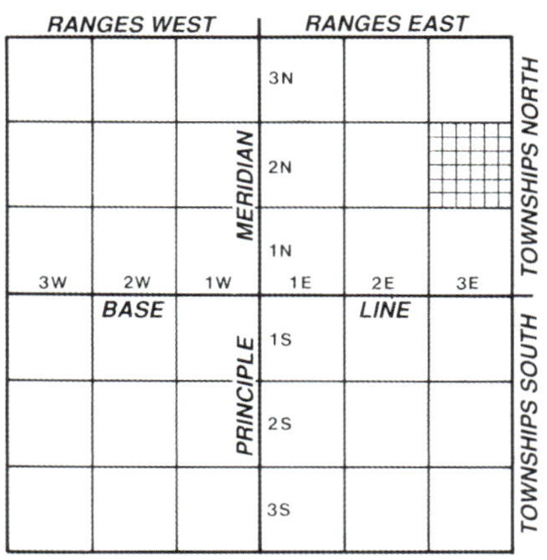

RANGES WEST | RANGES EAST; 3N, 2N, 1N, 1S, 2S, 3S; 3W, 2W, 1W, 1E, 2E, 3E; BASE LINE; PRINCIPLE MERIDIAN; TOWNSHIPS NORTH; TOWNSHIPS SOUTH

TOWNSHIP AND RANGE SYSTEM

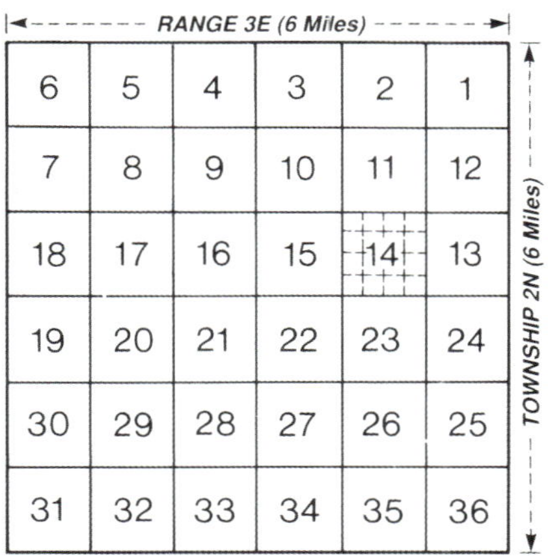

RANGE 3E (6 Miles) — TOWNSHIP 2N (6 Miles)

6	5	4	3	2	1
7	8	9	10	11	12
18	17	16	15	14	13
19	20	21	22	23	24
30	29	28	27	26	25
31	32	33	34	35	36

A TOWNSHIP DIVIDED INTO SECTIONS
T2N, R3E
(Most descriptions in this book only define sections within township and ranges)

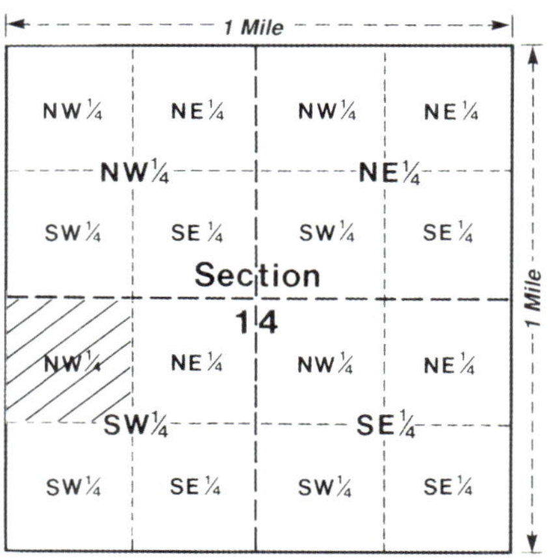

1 Mile — 1 Mile; Section 14; NW¼, NE¼, SW¼, SE¼

A SECTION SUBDIVIDED (640 ACRES)
NW1/4SW1/4 sec.14,T2N, R3E

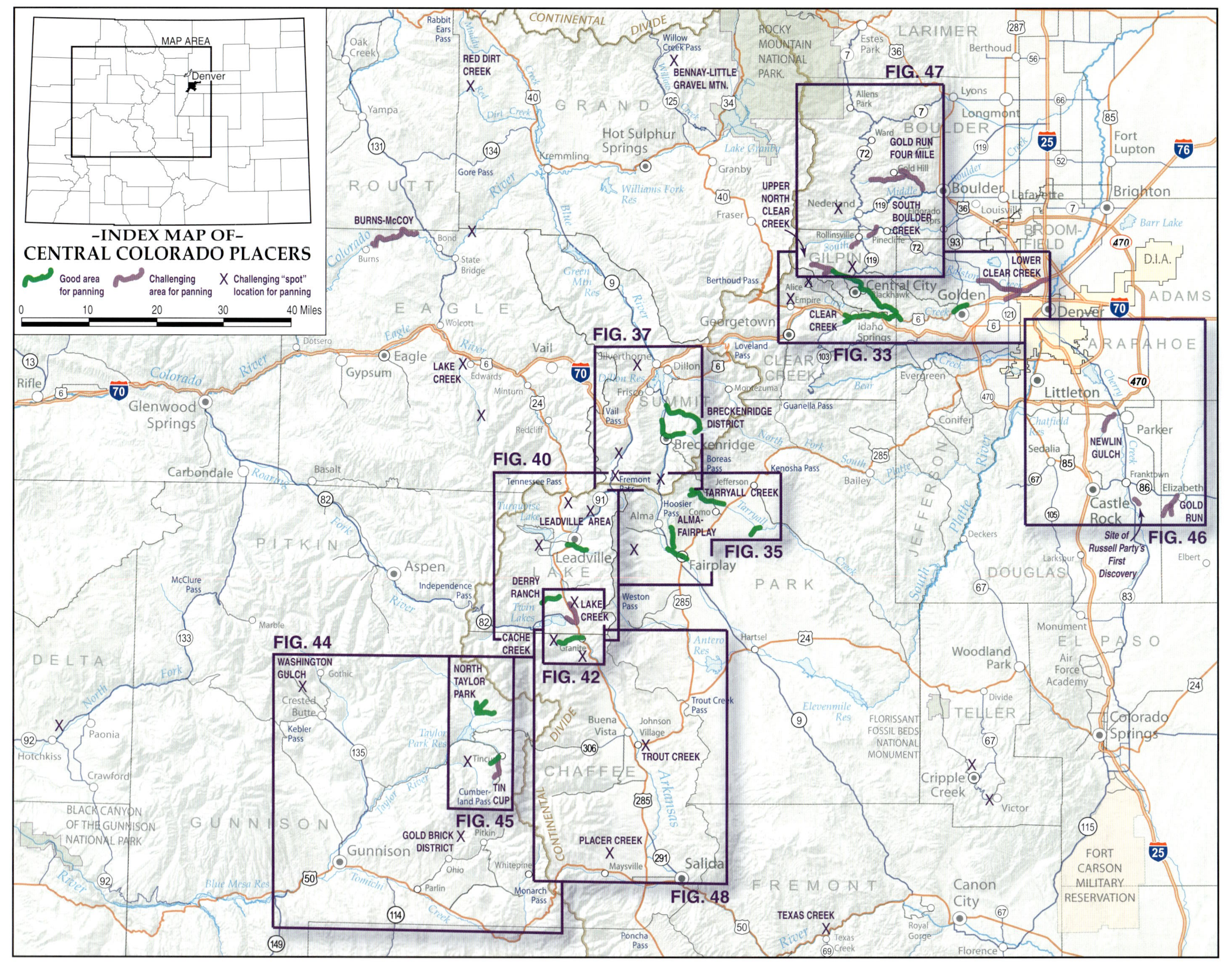

–INDEX MAP OF–
CENTRAL COLORADO PLACERS
Good area for panning
Challenging area for panning
X Challenging "spot" location for panning
0 10 20 30 40 Miles
MAP AREA
Denver
FIG. 33
FIG. 35
FIG. 37
FIG. 40
FIG. 42
FIG. 44
FIG. 45
FIG. 46
FIG. 47
FIG. 48
LOWER CLEAR CREEK
CLEAR CREEK
UPPER NORTH CLEAR CREEK
GOLD RUN FOUR MILE
SOUTH BOULDER CREEK
NEWLIN GULCH
GOLD RUN
Site of Russell Party's First Discovery
BRECKENRIDGE DISTRICT
TARRYALL CREEK
ALMA-FAIRPLAY
LEADVILLE AREA
LAKE CREEK
DERRY RANCH
CACHE CREEK
TROUT CREEK
PLACER CREEK
TEXAS CREEK
NORTH TAYLOR PARK
TIN CUP
WASHINGTON GULCH
GOLD BRICK DISTRICT
BURNS-McCOY
LAKE CREEK
RED DIRT CREEK
BENNAY-LITTLE GRAVEL MTN.
CONTINENTAL DIVIDE
ROCKY MOUNTAIN NATIONAL PARK
FORT CARSON MILITARY RESERVATION
FLORISSANT FOSSIL BEDS NATIONAL MONUMENT
BLACK CANYON OF THE GUNNISON NATIONAL PARK

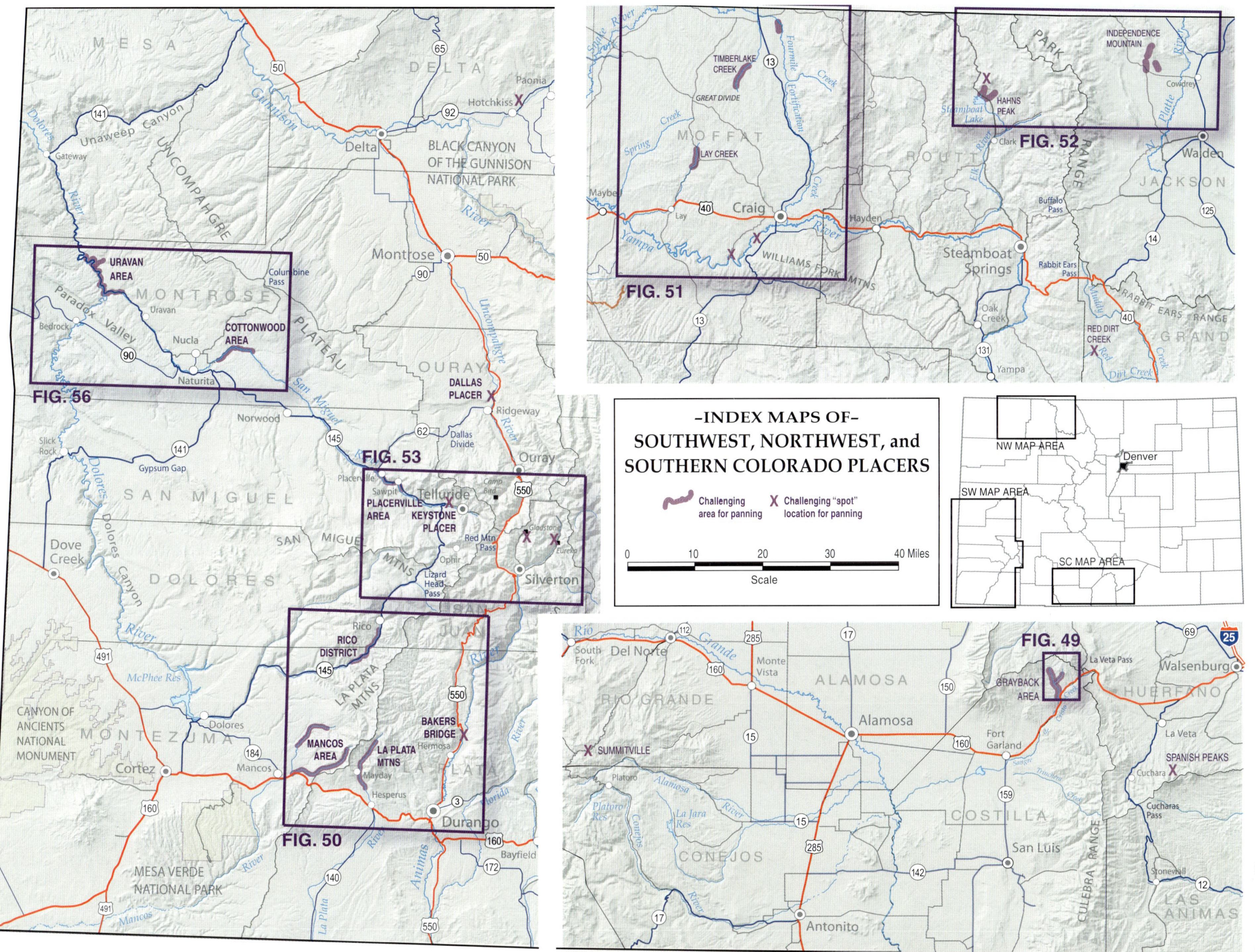
–INDEX MAPS OF–
SOUTHWEST, NORTHWEST, and
SOUTHERN COLORADO PLACERS
Challenging area for panning
Challenging "spot" location for panning
0
10
20
30
40 Miles
Scale
NW MAP AREA
SW MAP AREA
SC MAP AREA
Denver
FIG. 49
FIG. 50
FIG. 51
FIG. 52
FIG. 53
FIG. 56
INDEPENDENCE MOUNTAIN
HAHNS PEAK
RED DIRT CREEK
TIMBERLAKE CREEK
LAY CREEK
GREAT DIVIDE
URAVAN AREA
COTTONWOOD AREA
DALLAS PLACER
PLACERVILLE AREA
KEYSTONE PLACER
RICO DISTRICT
MANCOS AREA
LA PLATA MTNS
BAKERS BRIDGE
GRAYBACK AREA
SPANISH PEAKS
SUMMITVILLE
BLACK CANYON OF THE GUNNISON NATIONAL PARK
MESA VERDE NATIONAL PARK
CANYON OF ANCIENTS NATIONAL MONUMENT
Walden
Craig
Steamboat Springs
Hayden
Delta
Montrose
Ouray
Telluride
Silverton
Durango
Cortez
Dove Creek
Alamosa
Del Norte
Monte Vista
Walsenburg
Antonito
San Luis
Fort Garland
La Veta
La Veta Pass

INTRODUCTION

Gold panning is a great pastime. With some practice you can develop technique and pan cleanly and quickly. Panning often takes you to beautiful country and, if you wish, into remote country. And you find gold. It's great to see a trail of colors (fine gold) in the bottom of the pan! And it's fun to prospect.

There's always a chance of finding a rich pocket of gravel, or better yet, gold in bedrock. It doesn't happen often, but it has happened. The old time prospectors didn't find it all.

In 1932 or 1933 four miners placering on the East Mancos River discovered the Red Arrow deposit, a small but very high-grade free gold producer in Montezuma County. It was worked from 1933 to 1942 and yielded over 10,000 ounces of gold. Placering all the month before their discovery, the four men had produced only $17 worth of gold—a little more than one-eighth ounce for each man.

They were going more on hope than on encouragement!

About 1948 in early summer Irwin Corey and his father went to prospect Ice Lake Basin in San Juan County. They found no colors in the gravels of the stream that drains Ice Lake. When they were returning, Corey panned gravel from a brook near the South Mineral Creek camp ground and got one color. He followed the brook upstream and found more colors from place to place for about half a mile, where the brook headed in a spring. He prospected other gulches nearby and found a few colors on Clear Lake Creek. The colors continued a short distance up the creek and then up a tributary ravine.

From its head he went up the slope collecting samples, screening them and taking the fines down to water to pan. The colors eventually led him up a long, steep slope above timberline, where he found gold in bedrock near 12,600-feet elevation and later in some other outcrops nearby. He spent spare time during two summers following the colors, tracing the veins and staking his claims. His discoveries were the Burbank and Corey veins that the old timers somehow had missed.

In the summer of 1990 Shane Dodge found the "Turtle Nugget" at the Pennsylvania Mountain placer in Park County. This nugget—which weighs eight ounces troy—is on display at the Denver Museum of Nature and Science. It proves that there are some finds still left for us. A photograph of the nugget is on the front cover.

A number of placer deposits have no very obvious sources and certainly some free gold deposits remain to be found. Good luck!

SOME FUNDAMENTALS

Placers are deposits or accumulations of rock debris in which gold or other heavy minerals have been concentrated mechanically by natural processes. Placer gravel is the material that makes up a placer deposit without regard to its origin, size or shape. For instance, it may be stream deposits with any combination of sand, gravel and boulders; dune sand; beaches; weathered material lying on bedrock or even mill tailings.

The best-known placer mineral is gold; the most common placer bodies are gravels in stream beds. Stream placer deposits—alluvial placers—must have been man's first source of gold, and pans and sluices were his first tools for its recovery.

PANNING

Panning is a means of separating gold or other heavy density mineral fragments from rock waste material in which they occur, by shaking the material in a pan in such a way that the heavy minerals work their way to the bottom beneath the lighter common rock fragments. The material panned is usually stream or beach sand and gravel. Panning is a manual process done with the material to be panned soaked in water.

Panning is done wet for two reasons: First, the relative density of gold compared to that of gravel is about 1½ times greater in water than it is in air so that wet concentration is easier and more effective than dry. Second, in water the rock fragments are more easily displaced so the heavier material can sink downward more easily. (A more detailed explanation is given on p. 28)

Panning has two main purposes: sampling (prospecting) and cleanup. Panning is too slow for ordinary production. An experienced panner, if supplied gravel (and so does not need to dig it or pack it) and if that gravel is not clayey or cemented, can carefully pan up to 100 standard pans in a 10-hour day. This usually amounts to only 0.55 cubic yards of gravel.

PANS

Gold panning evolved world-wide since man first found and prized gold. A variety of styles of pans have evolved. In the United States the traditional pan is of steel, shaped as a frustum of a cone with a broad, smooth flat bottom and gently sloping sides, and may be smooth or have a few small parallel ridges that form concentric riffles about the pan. As sturdy plastics became available, other designs were developed, usually differing in that a part or all the rim has deep, specially designed riffles that the manufacturers claim "make panning easier and prevent spills of gold" or to save finer gold colors.

The pans come in a variety of sizes. The traditional, "standard" pan is 16-inches in diameter at the lip, the bottom is 10-inches diameter, the pan is 2½ inches deep and the sides slope 40 degrees to the bottom. This pan, level-full, holds about 20 pounds of gravel. Its pan factor (the number of level, or struck, pans of loose gravel equivalent to one cubic yard of gravel in place in the bank) ranges from 150 to 200 but is commonly about 180.

The standard pan is too big for convenience and the so-called "half-size" pan is better for general use. It is easier and less tiring to carry and easier to use in small streams. A skilled panner can wash a given amount of gravel about as rapidly with it as with the "standard" pan. The "half-size" pan has a lip diameter of 12 inches and a bottom diameter of seven inches, it is two inches deep and holds about nine pounds of gravel. Its pan factor is about 400.

Figure 1. Different kinds of pans–clockwise from left: green plastic pan with riffles, grizzly pan, standard pan, and "half size" pan. (Photo by Jack Rathbone)

Oils and grease that accumulate on steel pans from ordinary handling must be cleaned off before first using the pans and occasionally afterward. This is most easily done by "bluing" or "burning," which is passing the pan over an open flame such as that of a gas burner until the metal darkens or "blues," being careful not to heat the pan excessively. The dark color of the burnt or blued pan has the advantage of giving a good color contrast to any gold colors or fragments in the pan.

The plastic pans are available in a variety of colors said to show gold as well or better than a burnt pan. But they still must be cleaned from time to time, and the panner must be careful not to scratch them heavily.

NOTE:
Dimensions: In placer mining, gravel is usually measured in cubic yards and gold in troy ounces or in milligrams. A cubic yard of gravel in place ("bank measure") usually weighs approximately 3,000 pounds; one ounce troy equals 31,104 milligrams equals 1.097 ounces avoirdupois (avdp) (the commonly used system of weight). Also one ounce troy equals 20 pennyweights equals 480 grains. All references to ounces in this booklet are to ounces troy, all those to pounds are to pounds avoirdupois.
Prices: From the discovery of gold in Colorado in 1859 until 1934 the price of gold was about $21.00 per ounce, and from 1934 until 1968 it was $35.00 per ounce. Since then it has ranged up to $850.00 per ounce in 1980 and then down to $284.25 in 1985. All references to dollars in this booklet are based on the price of gold at whatever time is under consideration.

Recently a rectangular pan with flat bottom and sloping sides with ridges parallel to the bottom has been introduced. It is said to make removal of the coarser part of a gravel sample easier than with a conventional pan, but is probably less effective than the conventional pan in the final separation of gold from heavy sands.

In South America a pan commonly used, called a batea, is a pointed conical pan as deep or slightly deeper than a U. S. standard pan but broader (16- to 30-inches diameter). In southeast Asia and Indonesia similarly large shallow pans but with a rounded concave shape like a watch crystal are used. Miners accustomed to these pans can wash gravel as rapidly as those using the U. S. pan.

In Mexico, in a place with very little water in the Sonoran desert, I watched a prospector washing sand from which all coarse fragments had been carefully picked out. He washed this sand in a hornspoon, a spoon cut from a cow's horn with a bowl about the size and shape of a large kitchen spoon. He successfully recovered a few fine colors from a mine dump sample he had crushed that amounted to about half a cupful before sorting. (In most places you must wash one or even several pans of placer gravel—not half a cupful—to recover some colors.)

HOW TO PAN

Gold panning is like bicycle riding—easily learned but hard to describe. It is most readily learned by watching and panning beside an experienced panner, but following these suggestions you can teach yourself and soon begin to pan well:

1) If panning in a stream or lake, pan where the water is shallow but the pan can be held and shaken under water. If in a stream, the water must be only gently moving, so that the contents of the pan will not wash out. Ideally the water will be four to six inches deep, so that the pan can be rested on the bottom.

2) Panning is most comfortably done sitting on a rock in the stream, holding the pan between the knees. Some panners will hold their knees far apart with their arms and the pan between them. Other panners will hold one arm and the pan between their knees and one arm reaching under a knee to the pan. The first position is easier with a "half size" pan; the "standard" pan requires the second, especially if the panner is small. It is less comfortable to squat beside the stream, leaning over to pan. Knee-high rubber boots are a plus, and fishing waders are good for working and squatting in foot-deep water.

3) If panning away from a stream or lake, work over a washtub or other container of water large enough that the loaded pan can be submerged and, ideally, manipulated in it. This not only facilitates the panning but prevents having to pour water into the pan, permits filling the pan higher, and provides a trap for the tailings if desired.

Figure 2. This man is in the most comfortable position for panning and can continue working this way for some time. (Denver Public Library, Western History Department)

4) The pan should be filled about level full or a little less. Avoid heaping it with gravel. In some cases a sieving screen or "grizzly" can be used to eliminate coarse rock or woody material from entering the pan. Common screen sizes are one-quarter inch to one-half inch.

5) Throughout the panning operation you will be discarding portions of the material in the pan. You should watch this material and examine any fragment that has an unusual shape or color or is unusually heavy. With very rare good fortune you might find a nugget or a gold-bearing fragment of vein quartz. More probably you may encounter a cut nail or a metal, glass or porcelain artifact.

6) Submerge the pan in water and stir the gravel with your hands to wet it thoroughly. Wash any wood, plants, cobbles or large pebbles so that sand adhering to them drops into the pan and then discard them. If the gravel contains clay, work it so that the clay becomes suspended in the water and can be washed away. If the gravel contains clay lumps, these should be worked between the fingers until the clay "dissolves." Continue until the gravel is free of clay, the water in the pan remains clear as the pan is shaken and the wood, vegetation, cobbles and most of the pebbles have been removed.

7) Keeping the pan level, move it in strong twisting motions, clockwise and counterclockwise, so that all the gravel becomes suspended in water and the heavy materials can settle to the bottom of the pan. After, say, ten

motions remove the pebbles and granules from the surface. Repeat and continue until the larger fragments have been removed.*

Figure 3a. Step 6, submerging the pan full of gravel.

Figure 3b. Step 7, keeping pan level while washing gravel.

Figure 3c. Step 7 in footnote, author demonstrating use of grizzly insert. Note forward tip of pan while washing gravel.

Figure 3d. Step 9 in footnote, author demonstrating use of grizzly insert. Note forward tip of pan while washing gravel.

8) Alternating gentler side-to-side and circular motions, gradually tip the pan forward so that the surface of the gravel rises to the far-side lip of the pan. Continue agitation, letting some surface material wash over the edge, out of the pan. Level the pan again, repeat the twisting motions until heavy grains begin to appear at the outer edge of the pan. Again let some surface material wash over the edge. Continue leveling and tipping until heavy grains appear on surface near the pan lip. These grains are usually black, less commonly red, other colors or clear, but in any event different from the common sand grains in the gravel.

9) When dark, heavy sand grains appear, level the pan and shake it with side-to-side motions to suspend all the grains in water. While suspended, the dense grains can settle and the lighter grains of common rock minerals rise to the surface. Tip the pan forward until the gravel surface is at the lip. Gently raise and lower the pan in the water, tipping it forward slightly, so that the motion of the water rolling down washes the lighter grains on the surface out of the pan.

10) Continue this, from time to time leveling and shaking the pan again, until most of the grains of light minerals are gone and dark, heavy mineral grains begin

*Wells (p. 89) suggests that panning can be speeded by using a grizzly pan, a sieve made by drilling one-quarter inch holes in the bottom of a pan identical with the one used for panning. To use the sieve, place it inside the pan, fill with gravel and submerge in water in the usual way. When the material is thoroughly wet, lift the sieve slightly and twist it back and forth (under water) until all minus one-quarter-inch material has passed into the regular pan. Examine and card the plus one-quarter-inch material; wash the minus one-quarter-inch material in the usual way. Plastic grizzly pans are also available in specialty stores.

to spill out of the pan. At this stage some panners will lift the pan out of the water and dip it back in instead of just agitating the pan up and down in the water. If the common sand fragments are coarse or if there is little dark, heavy sand, combine a gentle forward and backward motion with the up and down movement to aid the water in moving the common sand fragments from the pan.

11) The pan now contains a heavy mineral concentrate with a few fragments and grains of tight, common rock minerals. Usually the amount of concentrate is small—a few spoonsful—and best cleaned as follows: remove the pan from the water, leaving only a small amount of water in the pan. Tip the pan so that the concentrate and the water lie in the angle between the pan bottom and side and gently swirl the pan so that the water moves in a circular path around the part. The water will wash the concentrate, and the movement should be so controlled that a concentrate "trail" forms along a part of the bottom edge of the pan. Doing this, the light minerals come to rest at the front of the "trail" while toward the back the minerals become progressively heavier.

Figure 3e. Step 11, a green plastic pan with riffles showing dark, heavy mineral concentrate and spectacular, bright gold grains. (Photos by Jack Rathbone)

Gold is heavy and, unless you have the very good luck to find a nugget, you will not see it until you have panned down the gravel to a heavy, usually black, concentrate. Fine yellowish fragments you may see before that stage are probably a variety of mica and almost certainly are not gold! Among the heavy minerals that work to the bottom of the pan, both gold and pyrite have yellow colors, but distinctly different ones. Most placer gold is a true yellow, becoming pale as the silver content increases, but pyrite is a brassy yellow. If you will hold a piece of gold jewelry (14 K or better) beside a piece of pyrite you will immediately see the difference in colors, and you are not likely to confuse them afterward.

HEAVY MINERALS

Front to back in the "trail" there may be grains of quartz and feldspar, then greenish epidote, then garnet (usually reddish) and black tourmaline, then brassy yellow pyrite and black magnetite and finally, with luck—gold. These are the most commonly found minerals in Colorado pan concentrates; usually only a few of them are present but occasionally still other heavy minerals occur.

The sorting of the minerals in the "trail" is only partial. It depends not only on the specific gravity of each mineral but also on the size of the fragments or grains. Small grains "ride forward," so fine gold colors may come to rest with coarser magnetite grains.

Occasionally the dark, heavy minerals are abundant. A single pan may have a half cupful of concentrate or more. The partner may decide to wash away a part of the heavy minerals by the method described above. If so, the panning must be done particularly carefully since gold colors cannot be separated from grains of heavy minerals such as magnetite as easily as they can be separated from common sand.

Since the most common heavy mineral is magnetite, many concentrates can be cleaned magnetically. The concentrates must be thoroughly dried so that the grains move freely. Pour a small amount of concentrate on a sheet of thin cardboard. Moving a small, strong horseshoe magnet, such as an ALNICO magnet, beneath the cardboard, draw the concentrate into a thin layer. Tipping the cardboard slightly, draw the magnetite grains upward leaving behind the gold and other non-magnetic heavy minerals. When some clean magnetite has been separated, remove it by covering the poles of the magnet with a sheet of paper and bringing it over, but not in contact with, the magnetite grains. The grains will "jump up" and adhere to the paper around the magnet and form a beard on the paper. Taking the magnet away from the cardboard, the magnetite can be disposed of by pulling the magnet away from the paper. If the magnet is not covered, the job of cleaning the magnetite grains from it can be very time-consuming.

After removing the magnetite, the gold can be separated with tweezers or picked up with a damp match stick.

Do not use mercury to pick up fine colors. Unless you do a very great deal of panning, the gold recovered is not worth the effort, the risk of handling mercury, or the cost. Mercury is poisonous, and the recovery of gold from amalgam by retorting is extremely dangerous.

If the concentrate contains many light-colored (and probably low density) grains, these can be separated from the heavy minerals by blowing. Again the concentrate should be dry so that it flows freely. Spread a thin layer of concentrate on a large sheet of paper and gently blow across it. (Your mouth should be as close to the paper surface as possible.) If the light grains can be readily blown beyond the heavy, dark grains they are probably of quartz or feldspar and can be brushed away. If they do not separate easily, they, too, are probably heavy minerals and should be kept in the concentrate and identified.

SALTING

Besides its uses in prospecting, production and cleanup, panning is often used for sampling. Salting is changing the apparent value of a sample, usually by adding a valuable material to it, but sometimes by "enriching" a sample site or by switching samples. Gold placers are usually sampled by panning or rocking samples. Since gold is valuable, spreading it over a sample site is not economical; most or all of the gold used needs to end up in the sample. Since placer gold samples are worked down to concentrates which are observed and weighed and no melting or chemical process is involved, salting must be done with particles of gold. This causes problems for the salter, because he should salt with real placer colors and needs a source of them. Some salting has been detected because the colors in the pan, under a magnifying glass, were seen to be cuttings and filings.

The traditional way to "sweeten" a sample was for the panner or person working the rocker to roll a cigarette, being careful to keep his hand over the pan or rocker. (Ideally the salter should be able to roll one-handed in a high wind) He had mixed a little gold with the tobacco in the Bull Durham bag, and he spilled a little of this tobacco into the pan or into the sample. This was very effective, since adding even one or two medium colors to a panful of gravel increases the value of the sample dramatically. (See pages 30 and 37 and remember that a standard pan holds only about one two-hundredth of a cubic yard in the bank)

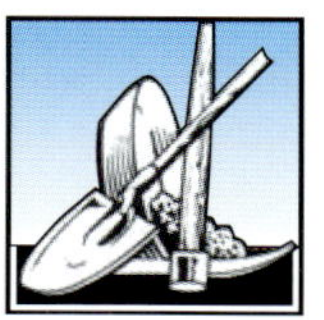

OTHER RECOVERY DEVICES

ROCKER

The rocker is a step upward in productivity. Even so, for some time it has been used largely for sampling; it makes high and reliable recoveries, but its throughput is low.

One man, washing samples, can rock, pan and make notes for eight 80-pound samples per day with a four-foot rocker, and in production-type work from one to three cubic yards can be washed per man shift.

A rocker somewhat resembles an infant's rocking crib. It is a sluice three to five feet long and 12 to 18 inches wide with a feeding device, mounted on cross rockers. It is usually made of sugar pine or redwood. The bottom of the sluice ideally is a mosaic of blocks with the grain running vertically, so that scraping it causes little damage. The sluice sits on a stout frame that has two cross rockers. At the upper end of the sluice a box (hopper) approximately 16 inches long and slightly less that the width of the sluice sits on rails across it. The bottom of the box is an iron plate punched with one-quarter-inch holes that occupy about 10 percent of its area. Beneath the box is a frame holding a slightly sagging canvas that slopes toward the head of the sluice. The canvas feeds the fines from the box to the head of the sluice and, sagging slightly, serves as a trap for heavy minerals. The sluice bottom commonly has riffles, but experienced operators can work rapidly and make good recoveries even without them. The rocker has a vertical, waist-high handle to shake it back and forth.

To use a rocker, set it up in a location convenient to the sources of both gravel and water, with the sluice inclined downward about 1½ inches per foot.

Shovel the gravel to be washed into the box and pour water over it. Shake the rocker with a vigorous motion, stopping after each stroke and permitting the box to slide and strike the side of the sluice. The fines pass through the punched plate. Add more gravel and wash until the gravel is exhausted, the box is full of coarse gravel or the riffles are full. When the coarse material remaining on the punched plate has been cleaned, examine it before throwing it away.

When the rocking is stopped, the material passing over the riffles settles and packs. This material must be loosened by scraping it backward, toward the head of the sluice, before resuming washing.

When the riffles are full or the washing is completed, wash the contents of the canvas apron and the material behind the riffles into a gold pan. After dumping the material on the canvas into the pan, clean the canvas and wash any material caught in it into the pan. Lift the riffles on the sluice and wash them and the floor and sides of the sluice and save the grains in the pan. Do the final concentration by panning.

Figure 4. Men cleaning placer concentrates with rocker. (Colorado Historical Society)

LONG TOM

The long tom is a sluice adapted to small-scale manual placer operations, much used in the early boom days. It consists of two sluice sections, one set above the other, with a continuous supply of water fed into the upper from a flume. The upper sluice box is narrower than the lower and has a grizzly (parallel bars or pipes serving as a screen) in its downstream end. The grizzly portion sits over the lower sluice.

Gravel, shoveled into the wider upstream portion of the upper sluice, is worked downstream in the water flow. The washed coarse cobbles are forked from the grizzly and discarded. The fines passing the grizzly wash through the lower riffled sluice where the heavy minerals are recovered.

SLUICES

A sluice for placering is a trough with rectangular cross section, usually fitted with riffles, through which a steady stream of water flows. Roughly sized placer gravels are washed in the sluice, carried along by the stream of

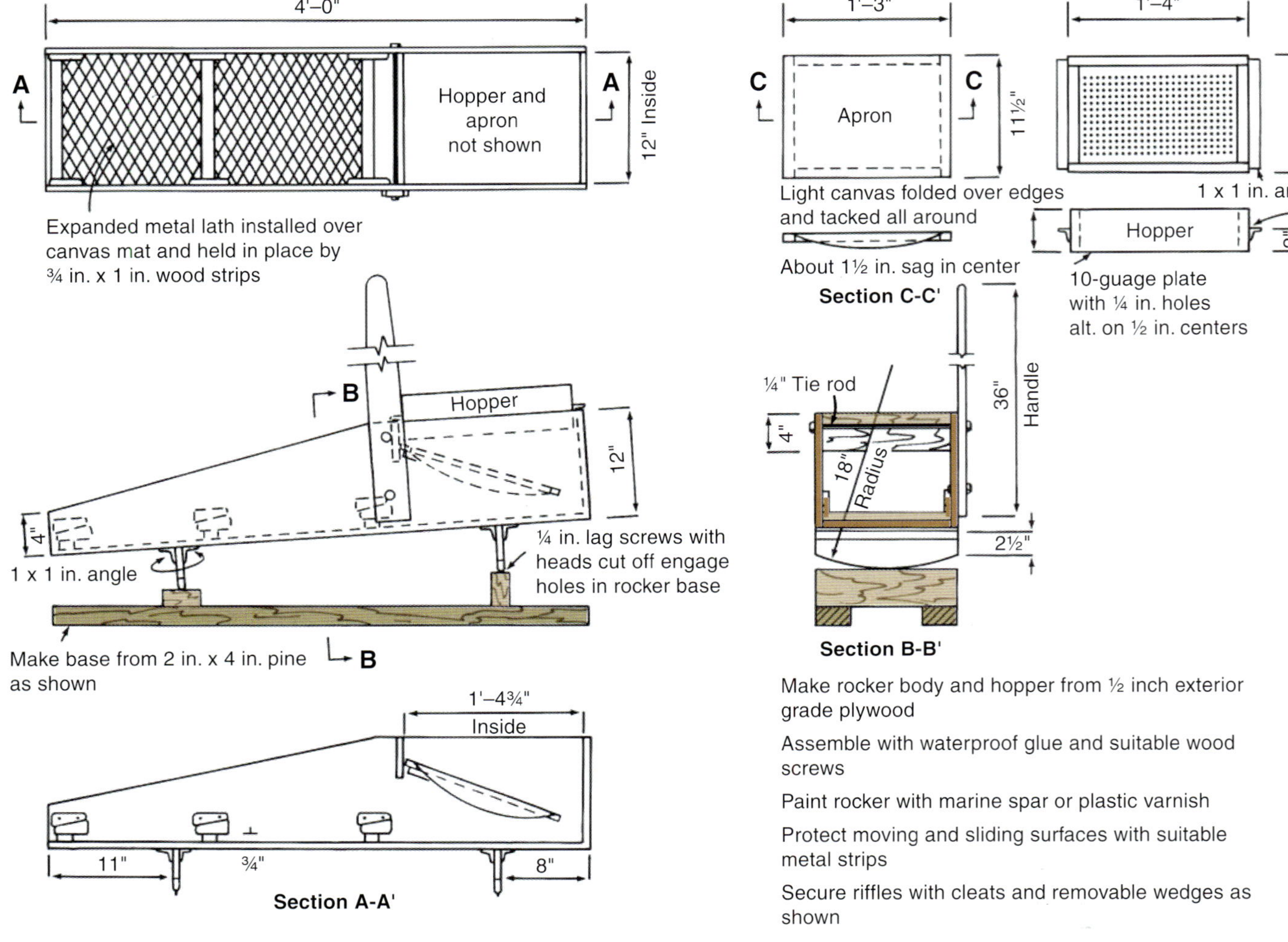

Figure 5. Rocker suitable for general use. With this outfit one man can usually wash the equivalent of eight 80- pound samples per day, including clean-ups, panning and note-taking. (After U.S. Bureau of Land Management, "Placer Examination, Principles and Practice," 1973, p. 78)

water, and the heavy minerals in them are trapped by the riffles. Sluices are usually built in sections, commonly 12 feet long, called sluice boxes.

Sluices must have a minimum slope to work effectively; it depends on the amount and size of the gravel washed, the size of the gold to be recovered and the amount of water available. Sluices commonly drop six to seven inches in each 12-foot long box. Where bedrock does not slope this much, sluices cannot be used very effectively; and below about 2½ inches in 12 feet they cannot be used at all. Where bedrock has only a minimum slope, sluices discharge their tailings at bedrock level and an elevator is required to stack tailings. Where bedrock has a slope appreciably greater than the minimum slope, the sluice can rise above bedrock downstream and tailings can be stacked directly.

RIFFLES

Riffles are barriers on the bottom of a channel that obstruct the flow of gravels being moved by water; these obstructions trap heavy mineral fragments contained in the gravels. Riffles do this in three ways: 1) by slowing material moving over them and giving it a chance to settle, 2) by forming cavities into which the heavy minerals can settle and 3) by causing eddies whose "boiling" action keeps heavy mineral sands from packing and has a classifying effect that concentrates the heavy mineral fragments. (Bedrock irregularities and boulders and cobbles in stream beds can have the same effect, forming so-called "natural riffles.")

There are many kinds of riffles used in sluices: poles, wood slats and blocks, rails and angle irons, metal mesh and even laid cobble stones. The riffles may run along or across the sluice. They are constructed so that they may be removed from the sluice during cleanups.

Two common types for small placers are "Hungarian" riffles and expanded metal lath. Hungarian riffles are transverse riffles, usually made of wood, often with a

Figure 6. Miners feeding gravel into sluice. Note the "shoveling boards," used so that the miners can easily throw gravel into the sluice boxes with their shovels. (Colorado Historical Society)

near-square cross section or slightly undercut on the downstream side to increase "boiling," spaced one to two times as far apart as they are high. Expanded metal lath is metal mesh with a diamond- shaped outline, stamped from a single sheet of metal, the mesh standing one-eighth to one-quarter inch high and sloping in one direction so "boiling" can be well developed. Since the mesh makes shallow, quickly- filled riffles, it is usually laid over burlap or cocoa matting sheets that serve as traps beneath the riffles. During cleanup the sheets are removed and washed to recover the gold trapped in them. Sheets of coarse, porous material frequently are placed beneath riffles of other types, as well, to increase their gold-trapping ability, particularly for fine gold.

The "golden fleece" mentioned in Greek legends was sheepskin riffle matting, loaded with gold. Sluices and riffles, like pans, have been used for literally thousands of years.

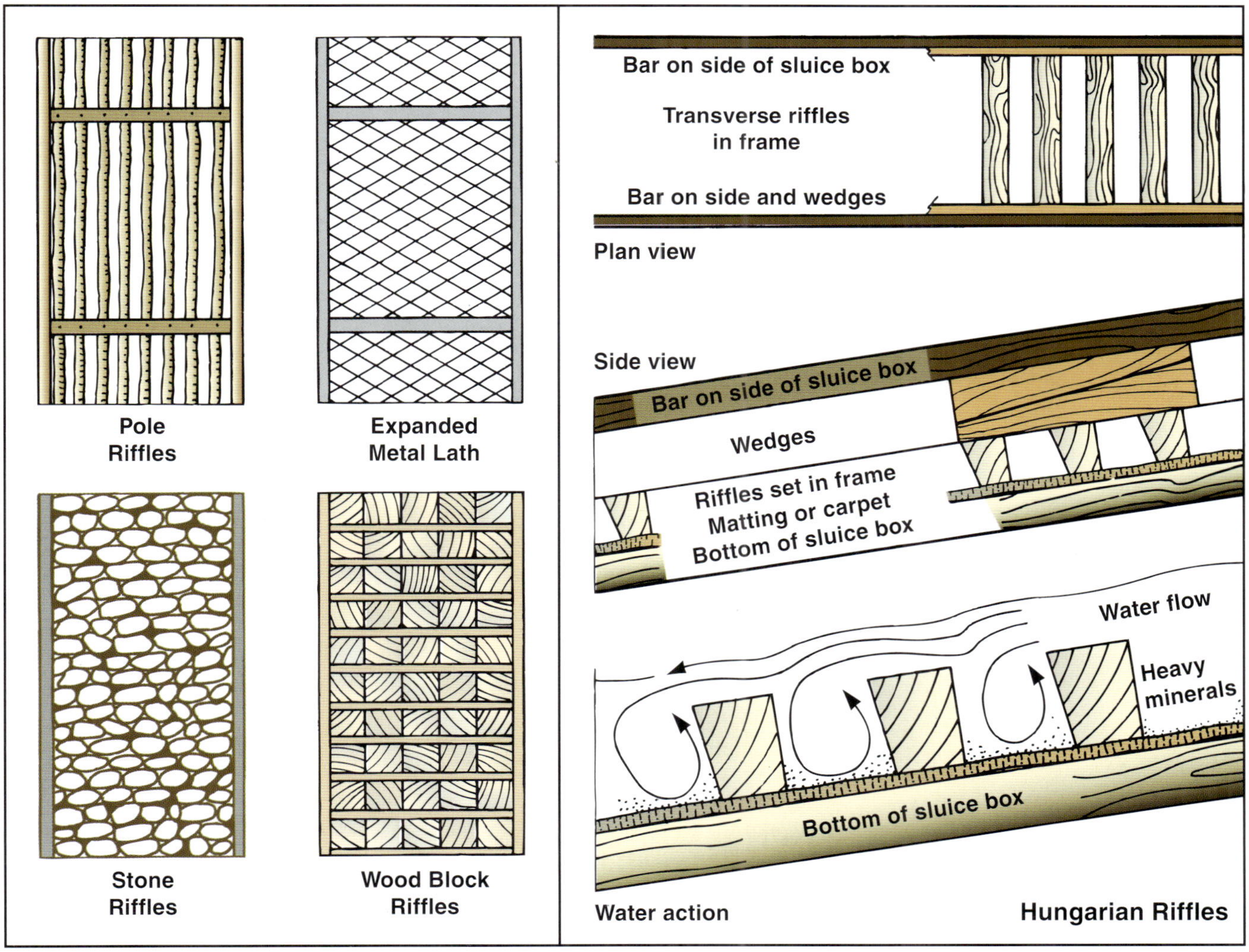

Pole Riffles
Expanded Metal Lath
Stone Riffles
Wood Block Riffles
Bar on side of sluice box
Transverse riffles in frame
Bar on side and wedges
Plan view
Side view
Bar on side of sluice box
Wedges
Riffles set in frame
Matting or carpet
Bottom of sluice box
Water flow
Heavy minerals
Bottom of sluice box
Water action
Hungarian Riffles

Figure 7. Types of riffles and water action over hungarian riffles. (After Peele, 1941)

DRY WASHERS

Panning is almost exclusively done wet. Some popular books refer to "dry panning" and to "dry panning contests," but since dry panning is so little seen and is unknown historically; probably it is very inefficient, if not ineffective. "Dry panning" suffers from the same difficulties as do almost all "dry washing" machines that have been invented; these are: 1) Classification is less effective in air than in water. 2) Because of the weaker classification, much closer sizing is necessary. 3) Because of the feebler classification and the need for closer sizing, the grains must flow freely—they must be dry. But most placer gravels, even those that appear driest, within a few inches of the surface become damp and will not flow. Also, "dry washing" processes cannot handle cemented or clayey gravels that can be broken down and washed in water.

POWERED WASHERS

There are some mechanical gold washers on the market small enough to be moved on a pickup or small trailer. One of the smaller and best known of these is the Denver Gold Saver. It consists of a feed hopper, a combined scrubber and trommel to wash and screen out one-eighth inch oversize material, a small vibrating riffled sluice, a water storage tank, and a centrifugal pump, all powered by a 1½ horsepower gasoline engine. The outfit weighs 750 pounds and has a rated capacity of two to three cubic yards per hour.

This and similar machines are too large for recreational placering and share a major disadvantage with the portable dredges mentioned below: They generate a substantial volume of fine tailings and often cause considerable pollution by muddying streams.

'PORTABLE DREDGES'

These units consist of nozzle, hose and pump that form a hydraulic sand elevator, a sluice device and a gasoline motor all mounted in a frame with enough floats so that the whole unit floats. They are used by divers with scuba gear to excavate and wash placer gravel in stream beds, gravels supposed in some cases not to have been reached by previous miners. Portable dredges cause serious pollution problems since they discharge their tailings directly into the streams in which they float. They are generally not suited to Colorado's placer streams, most of which are small and steep in comparison to those in Oregon and California where these dredges sometimes have been used successfully. A day's panning is backbreaking labor, and panning is too slow for regular gravel washing. Only rich gravels will pay day's wages now or would pay them even in discovery days. In those early times, very shortly after discovery of a deposit the miners would have built and begun using rockers, long toms and sluices. All these considerably increased the miner's productivity compared to panning. Nevertheless even the rockers and long toms had too low capacities, which became apparent when the rich discovery gravels began to run out. By the end of the first season the miners usually had begun to plan, if not to construct, ditches and facilities for methods with higher throughput.

PLACER MINING METHODS USED IN COLORADO

GROUND-SLUICING

Ground-sluicing is a placer mining method in which running water is used with manual picking to cut banks of placer gravel and to transport the fines and smaller stones to the sluice boxes. Only the fine gold is caught in the sluice boxes; the coarse gold lodges on bedrock where it is recovered by very careful cleaning.

Ground-sluicing is used where the gravel is thin enough that a person can safely work and pick down at the gravel face and expose bedrock, where the bedrock has sufficient slope that flowing water can move most of the gravel and where there is an adequate supply of flowing water. (If water can be made available under pressure, hydraulicking is much more productive.)

Ground-sluicing is begun by digging to bedrock a trench running in a downslope direction, ideally on one side of the deposit. A sluice is installed at the downstream end of the trench. Water is brought in a ditch to a high side of the deposit and turned to flow over the gravel bank into the trench. The trench is widened by picking and caving the gravel into the stream of water. The gravel is stripped to bedrock.

Using dams and walls the stream may be forced to flow against a bank to undercut it. A mining engineer writing in *Mining Engineer's Handbook* (p.10-541) commented, "Skillful use of dams forces a stream to do considerable work without supervision during a night shift."

Places where the bedrock is too low for the stream of water to clear it are cleaned manually later. Rocks too large for the stream of water to move are piled on bedrock in areas already cleaned. As far as possible the gravel beneath giant boulders is removed manually, but these boulders usually are not moved. The fines are swept to and along the sluice by flowing water.

Any coarse gold settles to bedrock, which becomes a natural sluice. Bedrock is carefully cleaned with shovels, hoes, hand scrapers, brushes and even with wires to clean crevices. Bedrock is often lifted, sometimes two or three feet deep, if sufficient gold has accumulated in fractures and cavities. Fine gold is washed to and recovered in the sluice boxes.

Two persons can move up to 30 cubic yards of gravel daily by ground-sluicing. This method is much more productive than the use of the pan, rocker or long tom alone. It also requires little investment to begin mining, if the ditch is not too long.

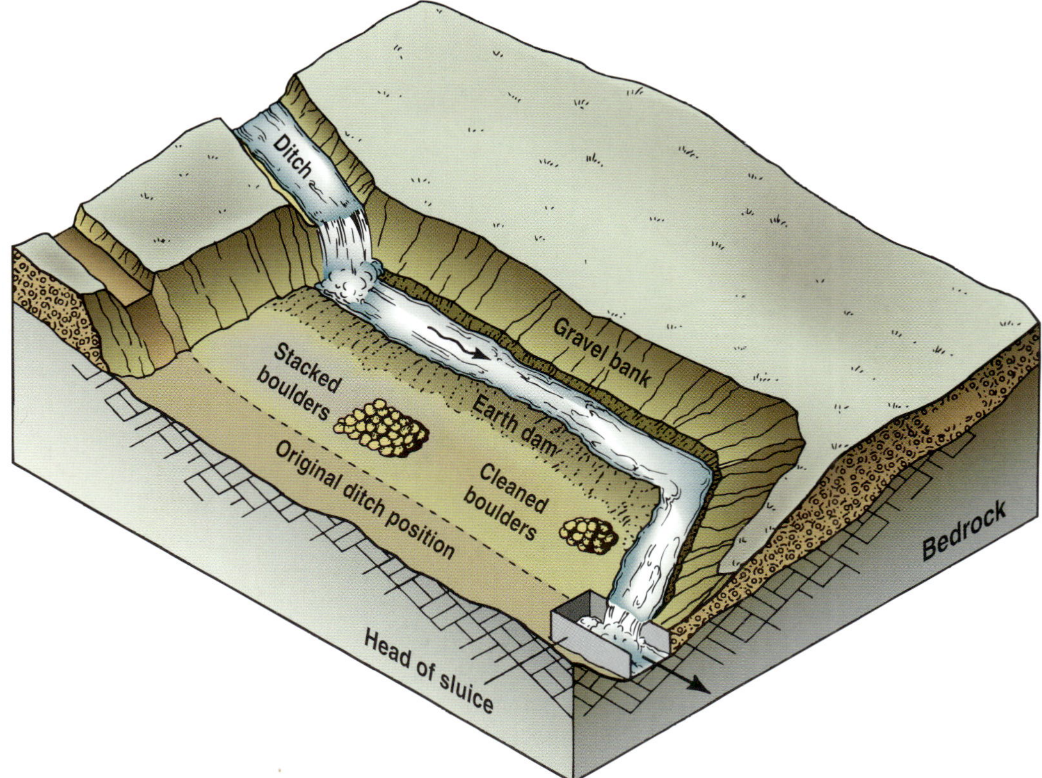

Figure 8. Layout for ground slucing. (Modified from Peele, 1941)

Figure 9. Miners ground sluicing with a monitor at Hahns Peak, Routt County. (Colorado University Historical Collections)

BOOMING

In places where and during times when water was scarce, the miners turned to booming, which is ground-sluicing done with surges of water. Dams were built in the ditch system, designed to be emptied in a flood. Since floods of water are more effective than smaller streams in undercutting banks, booming was also done intermittently to strip cover to expose gravel for later regular ground-sluicing.

Figure 10. These men are posed on a dam built for booming, Note the gate in the middle that permits very rapid discharge of the dam. (Colorado Historical Society)

SHOVELING-IN AND RELATED MECHANICAL METHODS

Shoveling-in is similar to ground-sluicing but, because of bedrock conditions or the amount of water available, the fines cannot be washed from the bank to the sluice but must be "shoveled in." Although the bedrock must still be carefully cleaned, much more coarse gold is recovered in the sluice boxes by shoveling-in as compared to ground sluicing. This is because the gravel just overlying bedrock is transported to the sluice boxes and then washed there. Since gravel can be shoveled economically only about 12 feet, the sluice line must be moved as excavation advances. Sluices can be fed either from their upstream ends or from the sides, depending upon the shape of the gravel deposit.

The method is adapted to shallow gravels above the water table. Average production is about five cubic yards per person per day.

Shoveling-in can be modified using scrapers or transporting devices and excavating machines of various kinds. When this is done, it is not necessary to move the sluice lines as often.

Figure 11. Miners are cleaning to bedrock in Illinois Gulch, Gilpin County—shoveling into the sluice the gravel originally resting on bedrock and any sand remaining from the sorting of gravel once lying higher above and now removed. The men are using the jet from a monitor and a stream of water to loosen the gravel and wash it toward the sluice. (Colorado Historical Society)

PORTABLE WASHING PLANTS

During the Depression and since, portable washing plants, or "doodle bugs," were often used at small placer mines. These consist of sizing and washing devices, sluices, jigs and conveyors and pumps for disposal of gravel tailings. The washing device is commonly a trammel with water spray. Jigs are tanks in which a pulsating water action is set up that has an effect similar to panning. The equipment is mounted on a single frame, skid-mounted or with wheels. Depending on the excavating or transporting device used, the plants often have conveyor loaders. These plants are used to work deposits above the water table, but with dragline excavation they can be used to work deposits below the water table.

Figure 12. Hallenbeck dry land washing plant and dragline, Derry Ranch, Lake County. (U.S.G.S. Photo Library, Capps, 1972)

HYDRAULIC MINING

Hydraulic mining is a placer mining method in which water under pressure, fed through a nozzle, cuts the placer gravel and transports all but the largest rocks to the sluice boxes. Bedrock still must be carefully cleaned, but usually most gold is recovered in the sluices.

The special, heavy nozzles used with their supports and fittings are called giants or monitors. They permit playing the water jet at many angles in nearly every direction. The jet can be played to cut the gravel bank, to wash gravel across bedrock into the sluices or to move large boulders on the bedrock floor.

Hydraulicking requires abundant water and topography suitable to deliver the water at high pressure. Large hydraulic mines with high banks, such as those at Cache Creek west of Granite and at Keystone west of Telluride, required water at heads greater than 200 feet. The water was delivered through long systems of ditches, flumes and tunnels to points above the placer site and dropped through penstocks to the mine and the giants. Cache Creek's highest bank is about 60 feet high; its ditch provided water at 300-feet head. Keystone's highest bank is more than 400 feet high and the mine's water head was 650 feet. (The pressure must be enough that the giant can be kept a safe distance from the gravel bank and still do effective work.)

The sluices were long. They recovered gold in their head stretches and carried all the material removed from the pit to a dump site and discharged it. A hydraulic mine's output is not determined by the giants' ability to cut gravel, but instead by the sluice's capacity to transport gravel with the water available.

In some situations the debris from hydraulicking can be piled on valley slopes, but the locations of the Cache Creek and Keystone mines near the floors of large valleys forced them to dump their tailings into those rivers. The Cache Creek mine was operated for 26 years, until stopped in 1910 by an injunction limiting hydraulic mining and controlling the introduction of sediment into the Arkansas River. The Keystone mine was closed in 1906 after five years' operations because of a landslide. Although it certainly had silted up the San Miguel River, no legal problems had yet arisen in that then isolated valley.

The Cache Creek and Keystone mines were large, but hydraulicking was combined with ground-sluicing in many small mines that had thinner gravels, to cut down banks instead of picking and washing them down. This not only increased productivity but permitted mining higher banks, These small mines worked with water heads as low as 35 to 50 feet.

Figure 13. A small hydraulic mine at Alma placer, Park County. Two streams of water cascaded over the cut wall moving the loosened gravel across the pit floor and through the sluice— the water from the monitor jets was not sufficient for this. On the left is part of the boom of the pit derrick, used to move boulders the jet stream could not roll. (Denver Public Library, Western History Department)

Figure 14. A small gulch mine employing a monitor for cutting the banks shows how in the space between the upstream bank and the head of the sluice the coarse rocks are removed by the miners and much of the fine fraction is washed into the sluice. The coarser gold will settle downward toward bedrock. Consequently, bedrock must be carefully cleaned before adding another box to the head of the sluice line. (Colorado Historical Society)

ELEVATOR PLANTS

From 1897 to 1906, especially in the Breckenridge district, hydraulic mining was attempted in the valley floors of large streams by using hydraulic elevators. The elevators could lift gravel and the water shot from the giants from pits excavated below the water table. A hydraulic elevator is a device in which a jet of water under high pressure, streaming up a steeply inclined large diameter pipe is made to lift up the discharge from a sluice at a low level and discharge it into sluices at a higher level. The elevator is fitted so that the discharge of the lower sluice, both water and gravel, is fed into the jet stream just above the nozzle at the bottom of the pipe.

Within the placer pits a monitor cut the banks downward to bedrock and washes the gravel into a bedrock sluice that carries it to the elevator. The bedrock sluice is for gravel transport only; no gold is recovered in it. The upper sluices, above ground level, are for both gold recovery and tailings discharge. Boulders too large to move in the bedrock sluice are set on bedrock out of the way or lifted out of the pit with derricks.

In order to work, elevator plants require great volumes of water under high pressure, bedrock less deep than the elevators' ability to lift the gravel, few large boulders and little black sand. Most of the elevator mining projects in Colorado failed because of inadequate prospecting. They encountered gravel values lower than anticipated, bedrock too deep, too many boulders and too much black sand.

Figure 15. The placer elevator plant at the Gold Pan pit on the Blue River just south of Breckenridge, Summit County. The pit was 73 feet deep to bedrock. At the pier on the far side three pairs of pipelines from the surface to the pit floor can be seen. One line of each pair is an elevator pipe, the other fed water under very high pressure to it. (Colorado Historical Society)

DREDGING

Dredging is the mining of placer gravels to beneath local water levels employing mechanical excavation and a floating plant. The dredges float in excavated ponds, called dredge ponds. Usually they extend beneath the natural water table, but sometimes they must be fed with water. Two methods of dredging have been employed in Colorado: bucket-line dredging, which employs a bucket-line excavator, washing plant and stacker mounted in a single hull, and dragline dredging, which employs draglines standing on dry land as excavators and a floating washing plant and stacker mounted on scows or pontoons.

Both types of dredging plant are very expensive, the bucketline much more than the dragline, and each must be designed for the deposits that they will work. To design the dredge and to plan the mining project both require that the exploration and testing have been correctly and adequately done. There must be sufficient gravel of high enough grade that it can be worked at a profit and repay the cost of the project. Depth to and nature of bedrock, number and size of boulders, the nature of the gravel and of the gold it contains must all be known to design the excavator and washing plant. Where dredging is practical, a dragline dredge will be employed instead of a bucket-line dredge if the reserves are not large enough to justify the much greater cost of the bucket-line dredge or if there is not enough room for it to maneuver. Draglines require easily excavated gravel with few boulders since they lack the digging power of the bucket-line. They also require soft, fairly smooth bedrock to get good cleanup. Draglines can work to a maximum depth below water level of about 30 feet; bucket line dredges can work much deeper.

BUCKET-LINE DREDGES

The bucket-line is exactly that, an endless chain made of links pinned together like a bicycle chain. Each link is a bucket with a separate lip piece, both made of manganese steel. The bucket-line is mounted on the digging ladder on tumblers at either end and rollers between them. The digging ladder is pivoted at its top to the top of the main gantry, which stands near the mid-point of the dredge. The digging ladder is raised and lowered by lines over the bow gantry powered by a winch on the deck. The bucket line is driven by the main power units through the hexagonal upper tumbler. The main hopper of the washing plant is set beneath the upper tumbler, so that the buckets discharge into it.

A dredge has a heavy steel spud collared at the stern and suspended from the stern gantry. The spud is heavy enough that its weight can force it into the placer gravel. Near the bow are the swing winches with lines anchored to the banks on either side of the dredge.

The dredge digs with the bucket-line, which excavates downward as the digging ladder is lowered. The dredge is turned on the spud by the swing lines, so it can excavate an arced shell from the bank as the ladder is lowered and raised, the dredge turned slightly and the ladder lowered and raised again. When the arc has been cut, the ladder is raised above the height of the bank, the spud is raised, the dredge is pulled forward with the swing winch, the spud and anchor lines are set and a new cut begun. Boulders too large to lift with the bucket-line are worked around and pushed to one side by turning the dredge and lowered ladder.

The washing plant has a trommel in which the gravel is washed and sized, usually to about one-half- inch diameter. The oversize goes to the stacker, a conveyor that extends behind the stern, and dumps the oversize into an already mined part of the dredge pond. Undersize is fed to the tables and to rougher jigs. The tables are groups of riffled sluices built side by side. Jigs are tanks in which the flowing water with the gravel fines is given a pulsating action through elastic tank walls. The pulsating action is similar in its effect to panning. Periodically, as heavy fragments settle in the lower part of the tank, called the hutch, they are drawn off through the bottom. The concentrate from the tables and rougher jigs is sent to the gold room. There it is washed in clean-up jigs, whose hutch product is treated in an amalgamator. The material produced in the amalgamator is retorted, then melted and gold is poured. The tailings from the tables and rougher jigs may flow over the tail sluice into the pond or may go over the stacker with the coarse gravel, depending on the nature of the original gravel and the ground recovery method being used.

Colorado's largest bucket-line dredge (called the South Platte No.1) (figs. 16, 17, and 18), was built below Fairplay on the South Platte River in 1941. Some details of this dredge are given in Table 1.

DRAGLINE DREDGES

A dragline dredge is like the dredge described above but does not have the bucket excavator or spud. A dragline standing on dry land feeds gravel into the feed hopper of the floating washing plant. As mining progresses, the dragline retreats and the dredge advances, leaving a tailings-filled pit behind it. The washing plants are similar to but smaller than the bucket-line washing plants because the capacities of draglines commonly, used are less than those of usual bucket-lines.

DRIFT MINING

Drift mining is underground development and production from buried placer deposits. The term is applied only to mining in deposits of unconsolidated or slightly cemented gravels, not to mining in strongly cemented deposits. The mines of the Rand in South Africa are not called drift mines, although they develop fossil placer deposits. In Colorado drift mining has been of three kinds: drift mining along the bedrock channels of valley placers, drift mining along channels of terrace placers buried beneath modern alluvium and drift mining along bedrock channels of ancient alluvial placers in Tertiary sedimentary rocks. The first has been done in many places but notably in North and South Clear Creek and in tributary gulches to the main streams in the Breckenridge district.

The second has been done along South Clear Creek and the San Miguel River near Placerville and the third in the Iron Springs and Cherry Creek Divide districts.

Figure 16. Bucket-line dredge elements of South Platte No. 1 dredge, Fairplay, Park County. The gravel washing plant, tables (sluices), and jigs, and the gold room are in the enclosed section of the dredge. The hopper and trommel are below the upper tumbler, the tables and jigs below them, so the gravel and slurries flow through the dredge by gravity. (Drawn from figure 17 photo.)

1. *Gangplank*
2. *Bow gantry*
3. *Digging ladder and bucket line*
4. *Upper tumbler and main drive*
5. *Tern gantry*
6. *Spud (hangs beneath these vertical lines)*
7. *Stacker*
8. *Tail sluice*
9. *Control room*

Figure 17. South Platte No. 1 dredge, Fairplay, Park County. (Colorado Historical Society, Zellers Collection)

Figure 18. Bow of the South Platte No. 1 dredge, each buckets' capacity was 13 cubic feet. (Colorado Historical Society, Zellers Collection)

Table 1. Description of South Platte No. 1 dredge

Type: Stacker and bucket-line	
Manufacturer: Yuba Manufacturing Co.	
Daily yardage, cu yd	
Average	13,000
Maximum	15,000
Depth dredged, ft	
Minimum	15
Average	85
Maximum	96
Hull	
Material	Steel
Length, ft	150.7
Width, ft	68
Depth, ft	11
Draft, ft	9
Digging ladder	
Size of buckets, cu ft	13
Number of buckets in line	
New	105
Minimum after use	98
Cutting speed, buckets per minute	
Average	28
Maximum	35
Screen	
Length, ft	54
Diameter, ft	9
Speed, rpm	10
Diameter of holes, in..	1/2 and 3/8
Gold saving plant	
Tables	32
Nugget sluices	2
Area of tables and sluices, sq ft	1,451
Jigs	
Rougher	16
Cleaner	2
Water pumped, gallons per minute	21,000
Total connected horsepower	
Maximum	2,400
Standby	1,100
Number of man shifts per day	
Direct dredge operation .	6
Total	18 to 24

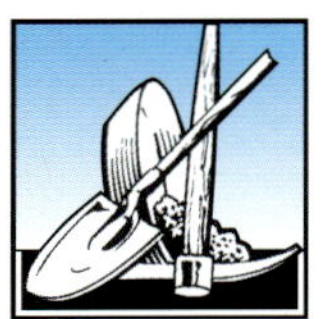

HISTORY

Colorado's "Pike's Peak or Bust" gold rush is usually dated 1859, but it began with discoveries of gold in 1858. However, there are reports of four gold discoveries before even that time in the area that was to become Colorado. In 1807 Zebulon Pike, for whom Pikes Peak was named, was a prisoner of Spanish authorities in Santa Fe. While there, he met an American trapper, James Pursley, who told him that he had found gold in 1804 near the headwaters of the South Platte. In 1848 members of the Fremont party are said to have found gold near Lake City, but the place has never been identified. In the period 1852-1858 soldiers stationed at Fort Massachusetts are said to have placered at Officers Bar in Costilla County. Just before this, in 1849 or 1850 two Georgians, John Beck and Louis Ralston, while on their way to California, found gold on Cherry and Ralston Creeks.

In California John Beck met William Green Russell, a fellow Georgian. They talked about Beck's early discovery and later corresponded and planned an expedition to Colorado. These two, with a party of 102 men, arrived at Cherry Creek May 20, 1858. They prospected Cherry Creek, Ralston Creek, lower Clear Creek and the South Platte River. They found gold, but not in paying quantities until they made a discovery in Russellville Gulch, in Douglas County. This was soon worked out.

During the early summer the party split several times until only Russell and 12 others remained in early July. The day after the final split the party found good placer gravel in two deposits on the South Platte, three miles above the mouth of Cherry Creek.

These two deposits were exhausted in about a month, but before that time passed a man named Cantrell, on his way to Kansas City, visited their camp and saw gold they had recovered. He took a sackful of placer gravel from their workings and panned it in Kansas City before witnesses. He obtained about 25 cents worth of gold. This was widely reported in the frontier papers and started the rush.

Russell's party prospected the streams at the foot of the mountains northward into Wyoming and returned to Cherry Creek in October. Russell and some of his men returned to Georgia for the winter, but a few men remained, they were joined by a party of prospectors from Lawrence, Kansas, and they wintered together beside Cherry Creek and the South Platte, not far from Colorado University's Auraria campus. Soon parties responding to the news of Cantrell's placer sample began to join them. Many of these emigrants, learning how small the discoveries had been, turned homeward, crossing out the "Pike's Peak or Bust" on their wagons and daubing in "Busted." Some people, however, decided to stay the winter with Russell's men and the Kansas party and the settlement that was to become Denver began.

January 7, 1859, George Jackson found gold at Jackson's Bar near Idaho Springs. During January the Deadwood Diggings on South Boulder Creek were found. The same month, or perhaps in late 1858, gold was found on Gold Run in Boulder County. May 6th John Gregory discovered the Gregory lode near Central City in Gilpin County. In June William Green Russell, who had returned from Georgia, found placers in Russell Gulch. Very quickly many bedrock gold deposits were found in Boulder, Clear Creek and Gilpin Counties and the development of Colorado's mining industry was assured.

Within a few months the sites of these discoveries were overcrowded, "the good ground was taken" and prospectors began roaming over Colorado's mountains. The placer deposits of Tarryall Creek and the South Platte River in South Park were found in July and August, 1859, those of Georgia Gulch and some others in the Breckenridge district in August and September. The placers of California Gulch, Cache Creek Park, McNulty Gulch and Tincup Gulch were found in 1860, as were shows of placer gold in Ouray and San Juan Counties. The Hahns Peak placers were discovered in 1864. Most of the State's placers, and all of the major ones, were found within a six year period.

The names of these, indeed of most of the placer camps, tell stories. Tarryall Creek was discovered by a small group of miners from the Front Range diggings. They formed a mining district and drew up rules for it. These rules apparently allowed more and larger claims than was usual in the very first camps. A later party coming to the area was disgusted to find all the ground taken, and this by few people. There was no reason to tarry at all. They went on, found gold on the South Platte, established customary rules and named their camp Fairplay.

The first panner of gravel from upper California Gulch reportedly said, "Boys! I've got all of California right in this 'ere pan," and the place was named. A party prospecting the Arkansas River valley turned into a small valley to cache some of their provisions. Panning gravel from some of the holes dug to bury their cache, they found gold and named Cache Creek Park. James Taylor washed the first gold from a gulch in a tin cup, according to the stories, and Tincup Gulch got its name. (Wouldn't that be tough: panning in a small, straight-sided cup?)

Stories grow, and maybe some of these I've repeated have grown, too. Here is one I've always liked, that probably has overreached. It was published in a mining company's annual report in 1910. A miner from Breckenridge in the discovery days was hunting near the present site of

Dillon. He sighted and shot a buck, tracked it to where it died and began to dress it. He noticed the buck's mouth and forgot the hunt. The buck's teeth had yellow specks over them! Gold! He backtracked the buck until he came to a spring where the buck had drunk, filled and washed a pan and got a nice string of colors. This deposit became the Saltlick placer. Isn't that greater than just saying, "The old timers found that deer came to a saltlick near the placer site?" Facts should never interfere with a good story!

Although some of the placers were worked for many years, the rich gravels of "boom times" were exhausted within three to five years in every camp. In this "boom times" period the rich deposits were worked with hand methods—pan, rocker, tong tom and shoveling-in—some ground as many as three times. Although the richest gravels yielded far more, typical yields per miner per day appear to have been $3 to $5.

In California in 1864 it was said that to wash a cubic yard of gravel by panning cost twenty dollars, to wash it with a rocker five dollars and with a long tom one dollar but to wash it by hydraulicking cost twenty cents. Costs in Colorado in the 1860s must have been much the same. Ground-sluicing was somewhat more expensive than hydraulicking. As technology evolved the costs of hydraulicking were substantially reduced.

Some of the deposits, when they no longer yielded prevailing wages, were worked by Chinese miners. Most of these men had migrated from China. They apparently were content with yields of $1.50 to $2.00 per day.

Figure 19. Chinese miners working at Tarryall district, Park County. Sizable groups of such miners worked at various times in Park, Clear Creek, and Gilpin Counties, but no photos of more than a few of them could be found. (David S. Digerness Collection)

Out of their meager earnings they often saved considerable gold and many sent their savings to China. Not all the men remained in this country. Many of them returned to China where they were able to live well on their pokes of gold. (A poke was the leather bag in which the early miner carried his gold.)

Some stories hinge on the Chinese miner. You may be told of a placer deposit with a recorded past production of thousands or hundreds of thousands "but the Chinese worked here and sent all their gold to China. If you count in all their gold, the production would be doubled." Remember that the Chinese arrived in the camps only when the boom days were over and commonly they only had ground that had been worked two or three times already. The gold they saved became a lot when it reached China, but in the American camps, even at the tail end of boom times, it was not reckoned so much. In most cases, the "lost" Chinese gold wouldn't change production figures greatly.

The Chinese miner was usually careful and slow. Ground he washed was washed. It has seldom paid to rework it. If you hear of placers worked by Chinese miners, look on a different property or in a different gulch for your gravel to pan!

After the boom times there followed a period of consolidation of small claims, construction of more ditches and flumes to bring water to the placers and of bedrock flumes to recover gold and to carry placer debris away. Properties were now worked by hydraulicking and ground sluicing and booming. This continued until the gravels remaining were too low grade to mine, until mining reached the level of the ditch and construction of a new, higher ditch was not justified (Gold Run at Breckenridge), until a disaster closed the mine (the landslide at the Keystone pit) or the mine was closed by injunction (Cache Creek Park mine).

Colorado's hydraulic mines were small in comparison to those of California. The largest were the Cache Creek Park mine in Chaffee County and the Keystone placer in San Juan County. The Cache Creek Park placer was worked for 58 years from 1860 to 1917, 48 years by hydraulicking. The placer was worked by two English companies in succession from 1883 until 1911, when an injunction stopped all hydraulicking activities. These companies appear to be the only foreign-controlled Colorado placer projects that were profitable, out of many floated overseas. The Keystone placer hydraulic operations began in 1901 and continued until 1906, when a landslide stopped operations.

During this period many ditches were built to supply water and permit booming and hydraulicking. Some were very impressive projects, especially remembering that only manual and animal labor was available for earthmoving. Early ditches were the Consolidated Ditch, 10 miles long, completed in 1860, that brought water from Fall River to the head of Russell Gulch and the

Cache Creek ditch, 16 miles long, completed in 1863, which delivered water from Lake Creek to Cache Creek Park. Many other ditches, some longer than these, were built later.

A spectacular flume was built on the Dolores River in the 1890s. It was part of the Montrose Placer Company's ditch that began a short distance below the Dolores' junction with the San Miguel in Montrose County. Depending on authorities, it was 10 or 13 miles long. (It was remote enough that facts did not have to tally. It still is!)

Approximately four miles of this project were in the Dolores River canyon where a timber flume was constructed, several-hundred feet above the canyon floor, pinned to the sheer rock walls and in one section even hung from walls arching above. Some remnants of the flume can still be seen projecting from the cliffs. This ditch and flume cost "something over $100,000." The ditch was completed in 1891, after a little over two years' construction, and hydraulic mining was begun. The project was abandoned, unsuccessful, after less than two years.

The Montrose Company's project and those of the other companies that actually operated on the lower San Miguel-Dolores River failed because of insufficient and/or technically improper sampling and also, probably, because of dishonest promotion. If there were any reserves of gravel of profitable grade in the two deposits, they were totally inadequate to even begin to repay the costs of the ditches and flumes and the other improvements needed to begin placering at either site.

On California Gulch in the early flush times miners complained about the "heavy blue sand" that clogged the riffles of the sluice boxes. It rapidly filled the riffles and forced frequent cleanups. When the cleanups were made, the separation of gold was more difficult. From 1860 until 1874, as the ground was worked and reworked and abandoned, the miners complained, but no one checked to learn what the "blue sand" was. In 1874 a hydraulic project was started to work gravels in lower California Gulch. Among the men involved were W. H. Stevens and A. B. Wood. Wood identified the "blue sand" as cerussite—lead carbonate -and, assaying it, found it contained silver. The two men prospected and in 1875 found ore bodies on both slopes of California Gulch above the placers. These were the first of the great lead-silver ore bodies to be found in the area and started the Leadville rush.

Not all of the placer discoveries were real. Some were fictitious, and one of these,—the Great Diamond Hoax—was famous in its time. In the Fall of 1871 two Kentuckians, John Slack and Philip Arnold, told stories in San Francisco of their discovery of gems in Arizona and produced a bag of stones to prove it. Their bag contained diamonds, rubies, emeralds and sapphires. (Until the end of the episode, after more than a year, no one remarked that these stones are formed in different ways and don't occur together in nature.) Soon a group of capitalists

Figure 20. The Montrose Placer Company's flume—the "Hanging Flume"— while under construction about 1890, Dolores River, Montrose County. (Denver Public Library, Western History Department)

Figure 21. The "Hanging Flume" after nearly a century, Dolores River, Montrose County. (Colorado Historical Society)

gathered, anxious to invest in the deposits. They included the Civil War generals George B. McClellan and Benjamin Butler, Charles Lewis Tiffany of Tiffany and Company and William C. Ralston, of the Bank of California, who with his associates had gained control of the older mines on the Comstock lode at Virginia City, Nevada, and manipulated their stocks to the group's benefit for many years.

Figure 22. U. S. Gold Placer flume under construction about 1888, San Miguel River, Montrose County. (Denver Public Library, Western History Department)

Slack said his discovery was in Arizona, but never said more. He made a number of trips, but always travelled on the Union Pacific railroad and got off or on at stations in Wyoming. Various people tried to follow him, but he always eluded them and always returned in too little time to have walked or ridden horseback to Arizona.

Slack and Arnold's first bag of gems, which they displayed in San Francisco, had been purchased with their savings of $15,000. (Slack had been an employee of a diamond drilling firm and had contacts to get crude diamonds and other stones.) Tiffany appraised them at ten times that much, perhaps because in his jewelry business he had dealt only with finished gems.

More gems were needed to salt the ground. Arnold went to London in July, 1871, and during that month bought more—with money the investors had advanced him. Then, in subsequent trips "to Arizona" from Wyoming perhaps in 1871 and certainly in 1872, Slack salted his site.

In May, 1872, Henry Janin, then the foremost mining engineer in the United States, was taken to the discovery. He was never told where he was, and was sworn to secrecy besides. Within an area of 160 acres, along a trail he found a mixture of gems in the sand, in anthills, on the surface at rocky outcrops. (Other accounts refer to an area a few hundred-feet across rather than a strip along a trail as having been salted,) As Janin worked, a surveyor staked the ground for the discoverers and the capitalists, but even he did not know the location of the claims he staked. On his return Janin called the discovery a fabulous find. After considerable publicity the San Francisco and New York Mining and Commercial Company was formed and its stock began to be promoted. The results were fantastic—for a few days.

Although Janin and the surveyor had never learned where they were, in the course of the publicity surrounding the discovery, they described the area they had seen—rolling country, no mountains to the north and east and so forth. Three geologists of the "Geologic Survey of the 40th Parallel" project (which was to become the sire of the U. S. Geological Survey) thought they recognized these descriptions. Their survey had covered a hundred miles on either side of the Union Pacific Railroad line, which Slack had always used on his "trips to Arizona."

The three set off southward from the Union Pacific in western Wyoming and soon arrived in an area that fit the description and where, after several days' search they found a diamond lying on an outcrop. They intensified their search and found many gems in the vicinity. But some things were wrong! Among them: They found a still more improbable mixture of gems; besides the diamonds, rubies, emeralds and sapphires there were garnets, spinets and amethysts. They found gems on rock surfaces and in cracks, but never within the rocks. They found gems in anthills, but only in anthills which had footprints around them and depressions in their tops, as if a stick had been pushed into them. There was an almost constant gem ratio—wherever they found a diamond, close nearby they would find twelve rubies. They concluded that the area had been salted.

When they returned to San Francisco, they announced their findings and the bubble burst. Janin was ruined professionally, Ralston's bank was embarrassed and weakened, Slack and Arnold disappeared. Clarence King, one of the three "detective" geologists, became well known and was made the first director of the U. S. Geological Survey, when it was formed a few years later.

The great gem fields were "on the edge of Vermilion Creek at its junction with Ruby Gulch at the foot of Diamond Peak" (Woodard, p. 33).

Many names have changed, but Diamond Peak is still shown on topographic maps. Its summit is in the extreme southeast part of sec. 32, T. 12 N., R. 102 W. Probably the gem fields were somewhere in a park called "Diamond Field," in secs. 14, 15, 22 and 23, T. 12 N., R. 102 W. There may be a few gems left there!

But back to gold and real placers. Many of Colorado's placer streams had rich, deep channel deposits on bedrock. Mining many of these began within a few years of discovery of each district, and channel gravels from drift mines contributed a large part of the "boom times" flush production. French Gulch had important drift mines in the early 1860s and North Clear Creek only a little later.

Figure 23. Two clam shell excavators built on site and used on the lower Blue River, Breckenridge district, Summit County, about 1900. These machines had capacities too small and slow to economically work the moderate grade gravels and could not move large boulders nor clean bedrock well. Note the wooden structures and the stacks indicating steam power. This was one of the methods attempted to work deep stream gravels before dredges were introduced. (Denver Public Library, Western History Department)

But there were some streams with channels below water level that could not be pumped or drained. The South Platte, the Blue and the Swan were the main ones. They tantalized miners for years.

From 1886 until 1905 many attempts were made to work these deep gravels with long bedrock sluices, with hydraulic elevators and with some unusual excavating and washing machines (see Figure 23). Only a few of these projects were successful.

In 1897 the construction of two dredges was begun on the Swan River. In operations in 1898 these dredges proved to be too light to deal with the large boulders encountered. In 1899 two heavier dredges were built at the same site and operated until 1905. In 1895 Ben Stanley Revett began to prospect the gravels on the Swan using churn drills, the same "cable tool rigs" used to drill many water wells and shallow oil wells in the U. S. until after World War II. This was one of the first occasions to use churn drills for this purpose, and Revett developed in the Breckenridge district many placer sampling techniques later used throughout the world. Within a few years the valleys of the Swan and the Blue and French Gulch had been extensively sampled.

In 1905 Revett built the Reliance dredge in French Gulch and it proved very profitable. The Breckenridge dredging industry began and from 1906 to 1939, except for 1938, there were one to five dredges operating in the district. The greatest activity was in 1918 and 1919, when five dredges were operating.

Dredges operated at many places in the State and at different times during the period 1900-1952. Many of the machines were small and probably largely "homemade" (on Placer Creek in Costilla County, Willow Creek in Taylor Park, Timberlake Creek at Iron Springs Divide, and Willow Creek at Hahns Peak). The largest was the South Platte No. 1 (see page 16) near Fairplay with a capacity of 15,000 cubic yards per day, built in 1940-1941, which operated 1941-1942 and 1945-1952. (Like other strictly gold producers, the dredge ceased operations during World War II, because of the "Gold Order.") This dredge was dismantled in 1980.

Figure 24. Churn drill being used for sampling placer gravels, Breckenridge district, Summit County, about 1900. (Denver Public Library, Western History Department)

The Depression times, with their massive unemployment and the increase in the price of gold from $20.67 to $35 per ounce on January 31, 1934, brought about a revival in placer mining in Colorado, which had almost ceased, except for the dredges in Breckenridge district. Many unemployed persons went out in the hills and attempted to rework abandoned placers. Some of the placers had become economic once again both because of the new gold price and because many of the unemployed were willing to work at almost anything legal, even if only poorly paid.

Golden had its "Placerville" where some 10 or 20 miners lived in shacks built of what scraps of material they could find. These men worked skim bars at the mouth of Clear Creek Canyon and perhaps made $1 a day (at the $35 price). Every town in the placer districts had similar groups of miners and many more lived out in the hills. Many of the miners quit in the first few years, for their living was very poor and other employment gradually became available. A few persisted in placering until the increase in business activity before World War II created many new jobs and prices rose. These men placered with pan, rocker and small sluice operation. Most of the Depression placer miners working on their own had hard lives rewarded with poverty; only a few were more successful.

Figure 25. Blevins dredge, Iron Springs Divide, Moffat County. Many of the early dredges apparently were designed locally and much of their machinery made locally. Note the small buckets in the bucket line. The placer gravel in the Iron Springs Divide district contained few fragments larger than pebbles, and this dredge worked satisfactorily several seasons. (U.S.G.S. Photo Library, Gale, H. S. 303)

Figure 26. A gold panning class November, 1934, on Cherry Creek, Denver County. A note on the photo is "Ways to make money by mining are being taught to scores of eager Denverites in the Opportunity School's mining classes financed by Federal Emergency Relief Administration funds. E. E. Miller, experienced mining man is the instructor. Both men and women attend as shown in one of the field classes in placer mining taken near the junction of Platte River and Cherry Creek. . . ." (Colorado Historical Society)

Few individuals and small partnerships could afford to mechanize their operations, but the increase in the gold price did bring about numerous larger-scale mechanized operations. Along with the price change, some other things differed from earlier times when placers had been actively worked. Earthmoving equipment was more generally available. Hydraulicking could not be used in many of the placer areas, except on the smallest scale, unless stream muddying could be controlled. Small movable dry land washing plants nicknamed "doodle bugs" and floating washing plants were developed. These plants, fed by draglines or power shovels, permitted working deposits too small or too irregularly shaped to dredge. There were a number of these in Colorado, particularly of the dry land type. Among them the placers in Clear Creek and North Clear Creek Canyons between Golden and Blackhawk were worked in several small washing plant operations, as were Cline Bench near Como and Pactolus near Rollinsville.

In 1922 the first gold-saving device was installed at a sand and gravel plant in Clear Creek valley in Adams County. Soon the devices had been installed in most of the plants in Clear Creek valley and on the South Platte River below Clear Creek in Adams and Jefferson Counties. After the South Platte dredge ceased operations in 1952 until 1974 when the last of the sand and gravel plants that had recovered gold in Adams and Jefferson counties ceased operations, nearly all the State's commercial placer gold production came from sand and gravel operations.

Figure 27. Women starring in a photograph of the same class, washing gravel on the Platte River with rocker and with inclined sluice and screen. (Colorado Historical Society)

ENVIRONMENTAL PROBLEMS OF PLACER MINING

From discovery times until the 1950s there were few attempts at surface rehabilitation after placer mining. At the gravel pits near Denver conversion of their surroundings to parks in which the pits remained as lakes was begun in the 1950s and was continued by the gravel companies and developers at many pits as they became exhausted and were closed.

Along the Blue River, dredge gravel tailings piles have been leveled and the areas leveled put to use from place to place. Of course, this is only a limited recovery. Many of the gravel tailings in Summit and Park Counties and elsewhere are valuable reserves of stone for construction aggregate. They probably will begin to be consumed in the not distant future.

Current practice in placer dredging and mining with dry land washing plants permits very good surface restoration. Before placering, the surface soil is stripped and saved apart. After mining, the coarse boulders and cobbles are returned first to the mined area and then the finer material is piled on them. (Older practice was just the contrary.) The tailings piles are then leveled as needed and the soil returned and planted. The method is very satisfactory.

I have seen photographs of areas in British Columbia before and a few years after their having been worked by a floating dredge. The dredged areas could scarcely be recognized.

Hydraulicking (hydraulic mining), especially large-scale hydraulicking, scarcely permits surface recovery. Large volumes of gravel are moved long distances in sluices and very large volumes of water are used. There are a few instances of mines in favorable locations where it has been possible to accumulate the tailings in hillslope tailings ponds and greatly reduce stream pollution, but the large gravel volumes and distances involved make return of the gravel to the cuts uneconomic and the separation of coarse and fine fractions almost impossible.

Surface placer recovery requires stripping and saving cover, constructing settling ponds and handling substantial volumes of gravel. The equipment required for this is incompatible with manual mining. Also, Colorado's placer gravels generally will not repay manual mining even without thought of surface restoration.

Recreational panning, if more than a few pansfull, can cause some problems! Be careful in digging not to bring down big banks. If you can separate the "pay" from the "lean," wash only the smaller volume of good gravel. If working a rocker, place it so the tailings are not discharged into a stream. Avoid "portable dredges." If you must use one, get permission from the landowner or Forest Ranger and shut down if your operations cause excessive turbidity that persists downstream.

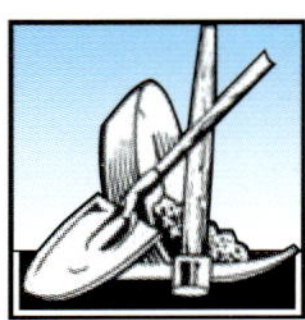

FUTURE OF PLACER MINING IN COLORADO

In Colorado there are few areas with potential placer reserves where the values of surface use do not exceed the potential values from placer mining. In many areas the costs of reclamation after mining would be prohibitive, if operating permits could be obtained. It is doubtful if any sizeable placer operations will be conducted in Colorado although some gold recovery may be initiated in conjunction with new sand and gravel washing operations.

Although the commercial future of placer mining in the State is bleak, placering as a pastime or hobby can be continued for many years. Good luck!

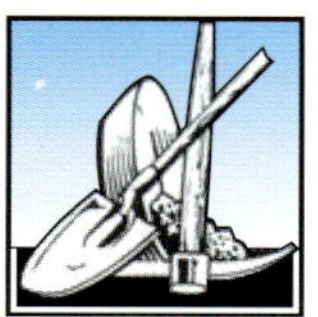

GEOLOGY OF PLACERS

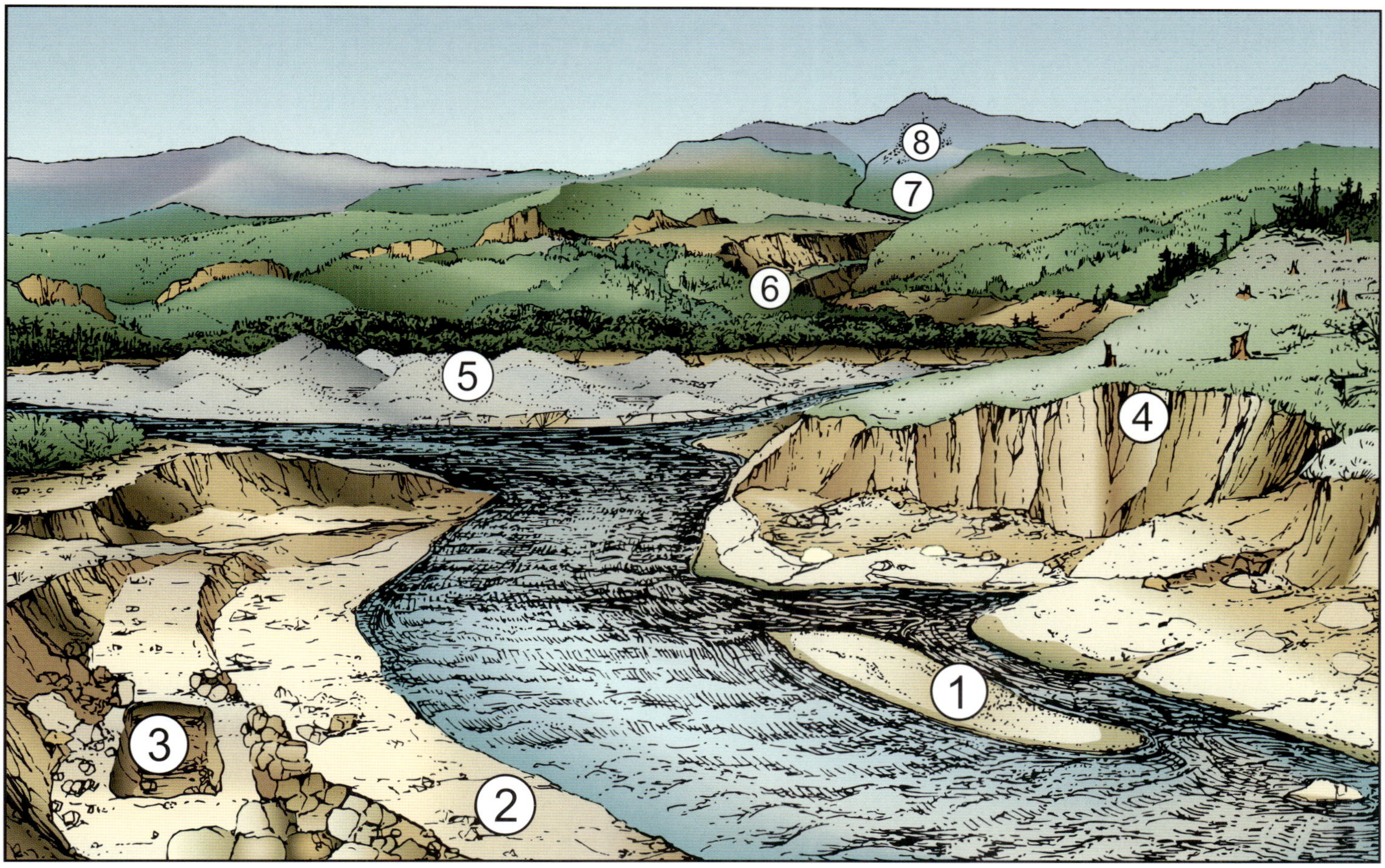

Figure 28. A hypothetical (idealized) landscape showing gold-bearing landforms.

1. *Skim-bar accretion deposits*
2. *Exposed stream gravels*
3. *Bench placer*
4. *Hydraulic cut in bank*
5. *Tailings piles and ponds in dredged area*
6. *Terrace gravel above river*
7. *Colluvial placer deposit*
8. *Bedrock gold source (lode)*

A placer is a deposit or accumulation of rock waste formed by processes of sedimentation, mass-wasting or weathering in which natural processes have mechanically brought about a relative concentration of gold or other heavy minerals. (Mass-wasting is the movement of rock waste in response to gravity—for example creep, mud flows or sliding.) Placer gravel is the material of a placer deposit, without regard to its origin or size. It may be stream alluvium (sand to boulders in any proportion), mantle material (soil and rock waste), glacial till, dune sand, or even mill tailings.

Gold is the best-known placer mineral, however placers are worked for other minerals such as cassiterite (an ore of tin), rutile (an ore of titanium), scheelite, ferberite and wolframite (ores of tungsten), platinum and diamonds. All have higher than average specific gravity and are relatively insoluble and resistant to alluvial processes. Colorado's placer production has been nearly all gold (with the varying amounts of silver naturally alloyed with it). Some tungsten minerals were once recovered in stream placers in the Boulder County tungsten district and silver ores were recovered from colluvial deposits near Georgetown in Clear Creek County.

Editor's note: Many prospectors believe that "gold is where you find it". You may occasionally find colors or trace amounts of placer gold in seemingly unlikely places. However, if you have a basic understanding of the geology of Colorado and the processes that form placer deposits you will not only find more gold, you will better understand why.

Placers can be classified according to the predominant process that formed them and their location (see Table 2).

Originally gold was formed in quartz veins or in fine disseminations in bedrock. The process of bedrock weathering liberates the gold fragments and forms eluvial (or residual) placers. As the weathering, erosion, and mass wasting continues the gold and other rock and mineral fragments move down slope to form colluvial placers. When the material comes under the influence of a stream, other moving water such as tides, or even wind, a process of density sorting begins and gold along with other heavy minerals are concentrated in alluvial placers.

It is favorable for the formation of gold placer deposits that the gold in the source areas be coarse, that the source areas be large with respect to the drainage area, and that there be a long period of deep weathering with a balance between transportation and deposition that favors concentration.

PLACER CONCENTRATION

Concentration of gold or other minerals in placer deposits comes about because of their high specific gravity. Placer gold has a specific gravity of about 19; common rocks and many rock-forming minerals have a specific gravity of about 2.7. If fragments of each move together in a stream of water or air, the gold fragments will tend to drift to the bottom of the stream and the rock fragments to rise above them.

Imagine fragments of gold and of rock, each of a size equal to the volume of one pound of water. In the air they will weigh, respectively, 19 and 2.7 pounds. The ratio of their weights is 19/2.7 = 7.0. When submerged in water, each fragment is buoyed up by the weight of the water it displaces—in this case one pound. Then the ratio of their weights becomes 19 - 1/2.7 -1 = 10.6.

So the tendency of gold to settle in preference to most rock waste is 10.6/7.0 = 1.5 times stronger in water than in air. This explains why concentration is more effective in water than in air. It is one reason why alluvial, or stream placers, are much more important than eolian, or wind-formed, placers.

PLACER SOURCES

Every placer deposit has an original or bedrock source. Eluvial placers lie upon their bedrock source; other types of placers are some distance removed from them. Some of the bedrock sources of a placer deposit may be, themselves, economic to mine; but others may be extensive areas of sub-economic mineralization. Examples of the second are the source area of the placers of Tarryall Creek, in Park County.

A cycle of erosion is a period of uplift followed by a period of erosion, If the formation of a placer is interrupted by uplift, this first placer is subsequently eroded and its gold may be transported and reconcentrated in a second, multi-cyclic placer. The first is the immediate source of the second, and both have the same bedrock source.

Table 2. Placer classification.

PROCESS	LOCATION	COLORADO EXAMPLE
Eluvial	Over bedrock source	Gregory vein, Gilpin Co. North Empire, Clear Creek Co.
Colluvial	Hillslopes	Slopes adjoining upper Gold Run Gulch, Breckenridge district, Summit Co.
Alluvial	Stream or valley	North Clear Creek, Gilpin Co. California Gulch, Lake Co.
	Bench or terrace	Idaho Springs, Clear Creek Co. Placerville, San Miguel Co.
	Fan	Terrace No. 2 and outwash plains, Tarryall Creek, Park Co. Buckeye Gulch, Lake Co.
Glacio-fluvial	Sub-glacial	Buckskin Creek and South Platte River, both above Alma, Park Co.
	Terminal or near-terminus moraine	Alma placer, Park Co., Snowstorm placer, Park Co.
	Outwash plain	South Platte River below Fairplay, Park Co.
Eolian		Great Sand Dunes, Alamosa and Saguache Co.
Marine	Beach	None
	Submerged	None
Fossil (any above)	(any above)	Cherry Creek Divide, Douglas Co. Iron Springs Divide, Moffat Co.

COLORADO MOUNTAIN LANDSCAPE FEATURES

Many of Colorado's placers are multi-cyclic, because of the development of the mountains' landscapes and of the glaciations the higher regions experienced. Throughout the State's mountains there are four frequent features: the highest peaks; mountain tops, shoulders and broader remnants that seem to have been parts of two groups of erosion surfaces, one well above the other, which slope away from the Continental Divide and the major range crests; and deep canyons beneath the lower group of surfaces.

These features can be seen particularly clearly from Interstate 70 at the head of Mount Vernon Canyon west of Denver. Looking westward from the viaduct you can see the snow covered ridge of the Continental Divide which dominates the far skyline. Some of the peaks are notably higher than the rest. Sloping downward toward the east from the lower parts of the Divide are some shoulders and mountain tops which appear once to have been part of the same general surface (Summit surfaces). Most of these are within a few miles of the Divide. Closer to the viaduct there are extensive remnants of a distinctly lower group of surfaces (sub-Summit surfaces) into which Clear Creek and its tributaries have cut deep canyons. We cannot see the bottoms of these canyons.

The high peaks are the few remnants of old mountains that were eroded, leading to the formation of the Summit surfaces. The shaping of the Summit surfaces probably ended in Eocene time, which was from 58 to 37 million years ago. Much later the Summit surfaces were uplifted and vigorous erosion began again.

The erosion which led to the formation of the sub-Summit surfaces probably concluded during Pliocene time, between 5 and 1.4 million years ago. Few, if any, bedrock gold deposits were exposed to erosion until that time. Consequently, the placers on Clear Creek and most elsewhere are not older than that. The deep canyons of the major streams which flow from the mountains onto the plains are the result of still another uplift and a much shorter time of erosion than those leading to the formation of the Summit and sub-Summit surfaces—the canyon-cutting cycle. The cutting of the canyons beneath the sub-Summit surfaces began during Pleistocene time; the canyons reached their present levels before the Illinoian glaciation, which began 302,000 years ago. In southwestern Colorado the cycle continued into Wisconsin time, which began about 132,000 years ago.

Colorado's known bedrock gold deposits did not extend upward to the levels of the Summit surfaces, although a few of them may have reached the levels of the lower sub-Summit surfaces. The only placer deposit which has been found on the Summit surfaces is that of Bennay Mountain-Little Gravel Mountain in Grand County. It and the fossil placers of Eocene age at Iron Springs Divide in Moffat County apparently derived their gold from mountains in Wyoming. The only placers known that were formed during the development of the sub-Summit surfaces are the fossil placers of Oligocene age at Cherry Creek Divide in Douglas and Elbert Coun-

Figure 29. The Front Range as seen from the viaduct at the head of Mt. Vernon Canyon on 1-70 west of Denver. The snow covered ridges of the Continental Divide on the skyline and most of the farthest tree-covered crests are remnants of the Summit surfaces. Remnants of the sub-Summit group are the lower tree-covered mountains in the far and middle distance. Deep canyons, such as Clear Creek's, which is just behind the rounded, tree- capped hill to the right, have been cut hundreds of feet beneath the low sub-Summit surfaces. Outside this photo are two peaks over 14,000 feet elevation that project above the Summit surfaces----Mt. Evans (to the left) and Longs Peak (to the right).

ties. These are thought by many geologists to have received their gold from the Central City region, but by some others to have been derived from a source in the Foothills to the southwest, since eroded away.

Most of Colorado's bedrock gold deposits began to be exposed during the canyon-cutting cycle. Placers have developed during that cycle and since, interrupted by pauses in the canyon-cutting cycle and affected by Pleistocene glaciations.

In general, the placer deposits in the San Juan mountains whose original sources were bedrock gold deposits are younger and less well developed than other Colorado placer deposits. Most of the San Juan gold-bearing ore deposits were well below the sub-Summit surfaces and began to outcrop later in the canyon-cutting cycle than deposits elsewhere. Consequently, the time for liberation and accumulation of placer gold was shorter here than in other parts of Colorado. In the San Juan Mountains, canyon-cutting continued into Wisconsin glacial times and the Wisconsin glaciations were more vigorous than elsewhere in the State. The rapid stream and glacial erosion, with little time for weathering, resulted in the delivery of finer gold to the streams. Furthermore, most of the main streams in the San Juans flow in long, narrow canyons where the vigorous Wisconsin meltwater streams could destroy previously formed placer deposits.

ALPINE GLACIATION IN COLORADO

Of the four Pleistocene glacial stages in the Colorado mountain area there is widespread evidence only of the last two, the Illinoian or Bull Lake and Wisconsin or Pinedale stages. There are a few isolated deposits of till of older age on the lower sub-Summit surfaces. None is gold-bearing. Some gold-bearing high-terrace gravel remnants in Clear Creek canyon may be pre-Illinoian in age.

The Illinoian and Wisconsin valley glaciers occupied the higher tributary valleys but only the upper parts of the valleys of the major streams. Of the two major glaciations; the earlier Illinoian Bull Lake stage was more extensive than the later Wisconsin Pinedale stage.

PLACER GOLD

SIZE

Placer gold ranges in size from nuggets to flour gold. The largest mass of gold found in Colorado weighed 160 ounces {11 pounds avdp (avoirdupois)}. Called "Tom's Baby," it was found on Farncomb Hill in the Breckenridge district in 1887. It is uncertain now whether "Tom's Baby" came from one of the extensive colluvial placers on the slopes of the hill or from one of the lode mines. Even larger, undisputable placer nuggets have been found in California and elsewhere. Colorado's next largest gold mass, and certainly a placer nugget, was found at the Banner placer, also in the Breckenridge district, and weighed 72 ounces (4 pounds, 15 ounces avdp). In 1990 an eight-ounce nugget was found at the Pennsylvania Mountain placer, where in 1937 a 12 ounce nugget had been found during placer operations.

Prospectors called vein material broken loose from bedrock and transported into the loose material on the slopes below its source "float." Float provides the gold of colluvial placers. The "Gold Boulder of Summitville" is a spectacular piece of float found in the Summitville district in 1975. Nearly 20 percent of this 108 pound avdp boulder is laced with native gold weighing 316 troy ounces. It was found on a slope in the vicinity of the Dexter adit, beside a small pine and with an eight inch pine seedling growing in it. Moss and lichen covered all but a third of the exposed part of the boulder. If anyone saw it before its discovery and recognition in 1975, they must have concluded, "The yellow material is pyrite—there is too much there to be gold," kicked it and walked on. The bedrock source of the boulder has not been found.

Tom's Baby, the Pennsylvania Mountain nuggets, and the Gold Boulder of Summitville are on display at the Denver Museum of Nature and Science.

From these giant chunks, placer gold ranges down to float, flood or flour gold. This gold, such as that found on the Snake River in Idaho, is so fine that one ounce of it may commonly contain 3,000,000 particles. With such fine particles, surface effects overcome the gold's high specific gravity; and if the particles are not well wetted, they can be made to float on water. These particles are carried long distances by turbulent waters and deposited in skim bars (accretion bars) after floods. This gold is difficult to pan and easily lost, hence the name float gold. Such fine gold can be the product of weathering of very fine gold in the bedrock source or of long ages and miles of abrasion and transportation in streams of coarser gold, or both. The very fine gold found in Colorado's placers may also be the product of glacial abrasion and scour or, especially in Clear Creek/North Clear Creek and in valleys which once held stamp mills, such as those of the gold-mining camps of the San Juan Mountains, of milling. Cement Creek in San Juan County is one of the latter.

A color is a fine particle of gold found in the pan after washing a sample of gravel. A widely used scheme of classifying colors in placer sampling is (Wells, p. 93):

No. 1 color	4—less than 10 mg
No. 2 color	1—less than 4 mg
No. 3 color	less than 1 mg

In this scheme colors weighing 10 milligram or more are nuggets. Samplers also speak of "colors too small to count" which must approach flour gold range. This term indicates that careful panning can recover some fragments this size and the eye can recognize them against a background of black sand or a blued pan.

As gold fragments move downslope in colluvial placers they are little changed from the size and shape in which they were liberated from bedrock. Once they reach a watercourse, as alluvial transport occurs they are reduced in size. This reduction, in terms of amount of reduction per mile, occurs rapidly at first (within 5 to 10 miles) until the average size may be 10 to 15 colors per milligram but much more slowly thereafter. In terms of time though, gold transportation and reduction in grain size must be very, very slow, as will be seen later.

SHAPE

Coarse placer gold may be angular, hackly, wire, shot or flake gold. In general, as a gold fragment moves along in a stream it is flattened and broken and the edges are rounded. Angular, hackly and wire gold is rapidly changed to smoothed shot and flake shapes. Nuggets often contain quartz fragments or other minerals from the bedrock deposit. As the nugget is transported, these fragments break and drop off. The nugget becomes more compact and purer. An example of the change in placer gold shape in some alluvial placers in Australia according to distance from its source is (N. H. Fisher quoted in Wells, p. 192):

Distance (mi)	Character of gold
5	Rough, nuggety
8	Small nuggety, water-worn
11	Fine-granular
25	Fine-scaly

In Colorado's streams typical distances are somewhat shorter. If coarse or rough gold occurs downstream in a section of predominately fine gold, it suggests either gold freshly broken from a rock moving downstream or another, nearby source of gold.

COMPOSITION

Placer gold also changes in composition with time and as it is transported. All placer gold is alloyed to some degree with silver and sometimes other metals. The placer gold produced in Colorado has ranged from 700 to nearly 900 fine and some gold from Douglas County has been an exceptional 990 fine. (Pure, or 24K, gold is 1000 fine.) The fineness varies from stream to stream because of their different bedrock sources. However, in a watershed with a single source area, along a course downstream for several miles the fineness will be found to gradually increase. This is because of the very slow solution of silver from the outer parts of the gold fragment. The smaller the gold particles, the more surface area an ounce of gold will have, and the greater the silver solution effect. Flour gold is purer than most nuggets. Older gold, as from a high bench or a fossil placer, will be purer than modern gold liberated from the same source.

ELUVIAL PLACERS

Placer gold is liberated from its bedrock source principally by weathering and occasionally by erosion. Weathering releases gold fragments in about their original shape and size. Fragments of gold released by erosion are reduced in size as the rock holding them is worn away. Weathering breaks up a rock by solution, by chemical processes such as hydration and oxidation that change mineral volume (and in that way fracture rocks) and by mechanical processes such as frost wedging and root wedging.

When weathering attacks a gold-bearing vein or other bedrock deposit, a portion of the vein minerals are removed in solution, and a few fragments may be removed by percolating ground water.

However, because gold is highly resistant to chemical weathering, most of the gold fragments tend to remain in place, gradually moving downward toward less weathered rock as the mantle material is removed. A placer deposit concentrated in place by weathering processes is an eluvial placer. Since these placers are related to individual bedrock sources, eluvial placers are richest and most extensive where those bedrock sources are rich and extensive.

COLLUVIAL PLACERS

Eventually soil mantle material including weathered vein matter will be moved downslope by mass-wasting processes until it reaches the nearest stream course. Moisture in the soil mantle facilitates and hastens this movement. Flowing water moves principally over the surface of the soil mantle in sheet flood with no fixed course and in rill and gully wash. Lighter rock waste fragments are washed away at the surface but there is little downward concentration of the heavier minerals. Downslope movement of the soil mantle (with material from upslope sliding over material below), accentuated by processes such as frost heave, tends to lift heavy minerals away from bedrock. This is seen in the Pennsylvania Mountain placer in Park County, where in much of the deposit the best values were near the surface and the gravel on bedrock was barren. Placers such as these, which have been transported by mass-wasting only, are colluvial placers.

ALLUVIAL PLACERS

An alluvial placer is one in which transportation, concentration and deposition of placer gravel by running water have been the dominant processes forming the deposit. These are the most important gold placers in Colorado and in the world. Alluvial placers may be beneath present-day streams and their valley floors, beneath gravel terraces above the level of the valley floors or in alluvial fans, the cone-shaped accumulations of debris formed where steep, mountain streams flow onto gently sloping lowlands. According to their settings, they are called valley or stream placers, terrace or bench placers and fan placers.

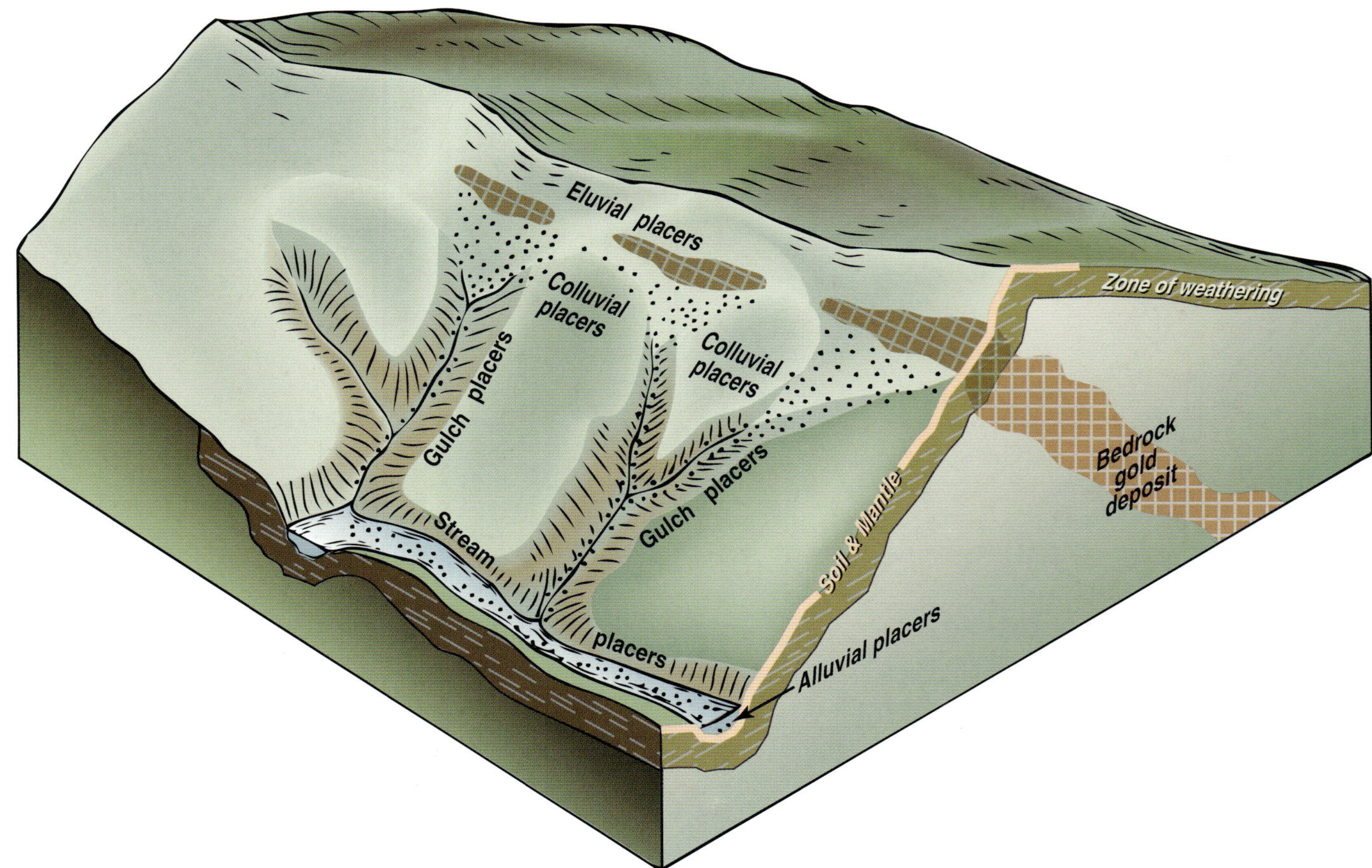

Figure 30. Eluvial, colluvial and alluvial placers.

ALLUVIAL TRANSPORTATION

A stream transports placer gold coarser than flour gold predominantly in its bedload. When gold approaches flour gold in size, the stream may carry it for long distances in suspension, so long as the water remains sufficiently turbulent. The bedload is the material of any size—sand to boulders—that a stream moves along its bed by rolling, sliding or saltation (skipping).

Most of the time, in any stream, at most places along its course, a layer of gravel separates the stream-bed from bedrock. At a given point on the stream the thickness of this gravel layer and the size and amount of the material moving past change with the speed of the stream. During a year the streams of spring runoff and those following cloudbursts are strongest; they will move larger fragments than will the streams at other times. Along a stream's course at a few places the entire gravel column will be in motion from time to time when these strong streams flow; but at many other places the gravel will only move in response to exceptional storms—the so-called "hundred year storms" or to extreme changes in annual stream flow.

At the Fairplay placer in Park County the gravels have not been moved down to the false bedrock since the beginning of the retreat in the Wisconsin glacial stage (Pinedale) about 23,000 years ago nor to true bedrock since the beginning of the Illinoian glacial stage (Bull Lake), some 302,000 years ago.

GOLD COMMINUTION

Just as streams transport gold fragments, they also comminute them, flattening, bending and breaking them as the gold fragments are caught between colliding, rolling or sliding rocks. Both a stream's comminuting gold fragments and its transporting them require very long intervals of time, and as the fragments become smaller their comminution becomes slower although their transportation becomes easier.

ALLUVIAL PLACER GOLD DEPOSITION

BEDLOAD GOLD

A placer gold fragment moves along a streambed until the current wanes and can no longer move it or until it encounters a trap. The trap may be a cobble or boulder which causes eddies and local spots of low velocity, a riffle or some natural object projecting slightly above the bed, an opening between fragments in the bed or a

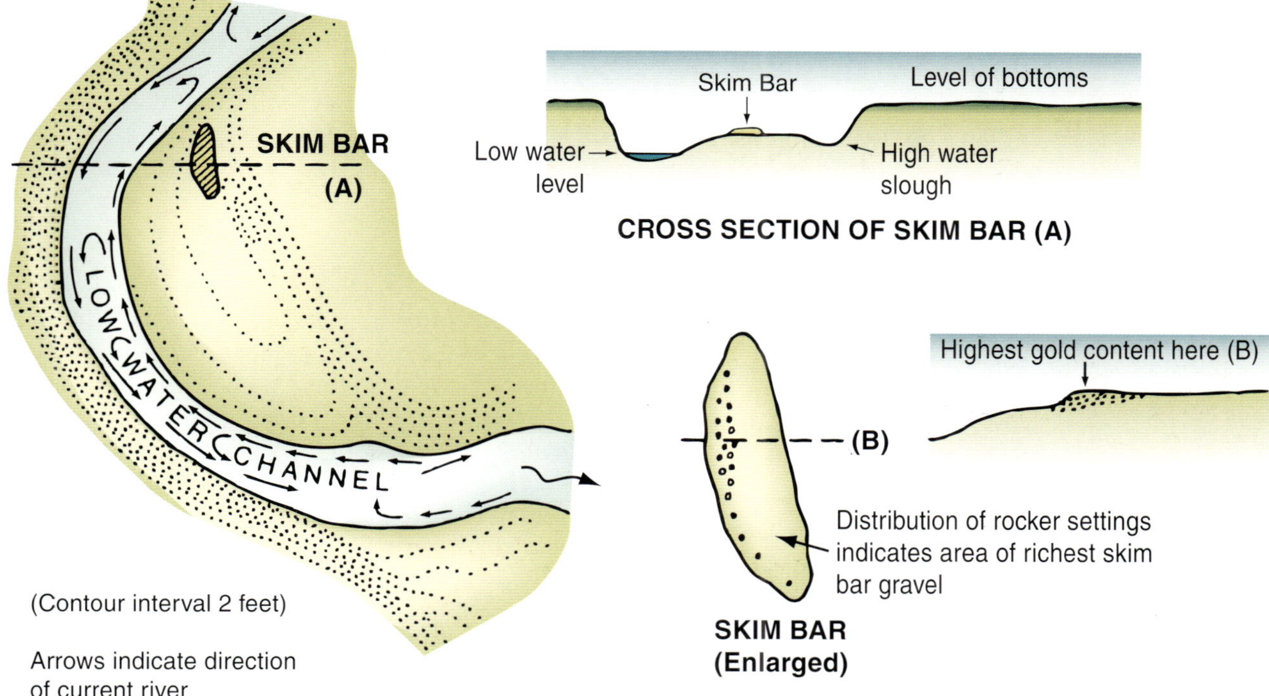

Figure 31. Location of flood gold accumulations on accretion or "skim" bars. (After Bureau of Land Management, "Placer Examination, Principles and Practice", 1973, p.16, and U.S. Geological Survey Bulletin 620)

bedrock crevice. (An old placer miner told me once, "You find gold where you find trout", and that seems to be a pretty good rule for streambed gold!) Because of its greater relative weight, the gold fragment will drop into the trap in preference to rock fragments which the current can more readily sweep along. As the bed is stirred, the gold fragment moves downward by stages into it, and a progressively greater current is required to move it each time. This continues until the fragment reaches bedrock, and even bedrock will be penetrated if it has cavities or fractures. If the gold fragment becomes trapped in bedrock, it cannot move again until the bedrock is eroded to free it.

SKIM BARS–SUSPENSION LOAD

Skim bars, which are mentioned elsewhere (page 34), have always attracted the placer miner's attention because they appear to be renewed each spring or after major storms. (Skim bars are accumulations of flour gold, usually on the upstream points of accretion bars and only in the top few inches of gravel, deposited by receding waters after floods. During the floods the fine gold is transported in suspension; it settles downward when the floods ebb. Accretion bars are low-level sand and gravel bars which build up gradually in streams, typically on the inner side of curves.) Some miners have deluded themselves—and some promoters have deluded others—on seeing this float gold washed in each year, into believing that gold grows in placer mines and that an exhausted placer in time will be replenished naturally. (The time I've heard was 100 years.) Don't you believe it!

For a given stream regime, on average, as much suspended flour gold is washed away when the floods begin as is dropped when the floods recede. In the bedload gold moves slowly, in terms of geologic time rather than human lifetimes.

GOLD TRAPS

Coarse Gold

Placer gold accumulates along stream channels in places where the velocity of the stream decreases rapidly. The sites of accumulation of coarse gold, transported in the bed load of the streams, are determined by bed and bedrock (or false bedrock) features. Those of float gold, transported in suspension high above the stream-bed, are determined by near-surface stream velocities.

Considering a stream in long section, neither the bed nor the bedrock (or false bedrock) of a mountain stream has an even slope. They are irregular, whether viewed broadly or in detail. This is also true of more mature streams on the plains near the mountain front, although their bed surfaces are less irregular. These irregularities form placer traps for coarse gold. Many of the smaller-scale traps have already been mentioned.

On the bed there are openings between fragments. Gold trapped in them moves slowly downward through them, and not all gold is on bedrock. In Colorado in thick gravels roughly 90 percent of the gold lies within three to six feet of bedrock or false bedrock, and usually in thin gravels most of the gold is in the bottom half of the gravel section.

On the bed or on bare bedrock cobbles and boulders cause eddies with local low water velocity. Gold may accumulate on the lee side behind and underneath cobbles and boulders. It then begins its gradual movement towards bedrock. Gold accumulates on the downstream side of natural riffles, just as it does in riffled sluices. Natural riffles may be dikes or layers of resistant rocks between softer ones or a contact of hard rock with soft. In either case, if the downstream side is abruptly lower than the upper one, bedrock gold tends to accumulate at the "step." Occasionally the resistant bands are visible on the adjoining hill slopes and the position of the step can be projected to the bedrock. Sometimes an individual boulder or cobble projecting slightly above the streambed will form a natural riffle as well.

On bare bedrock gold accumulates in cavities and in cracks. Potholes are often exceptional traps. Gold fragments accumulate in cracks in bedrock large enough to admit them. The depths to which they can penetrate is surprising. In some areas miners have found it worthwhile to lift as much as six feet of bedrock to recover gold. In others with less favorable bedrock there has been little or no penetration. The cracks may have developed from earlier fracturing or from weathering of platy rocks such as gneiss, schist, slate or shale. These rocks, particularly in areas of deep weathering, often develop excellent traps. The trapping effect of the platy rocks is more notable where their lamination is more nearly perpendicular to the stream course.

Several channels may develop on bedrock beneath a wide valley floor. The slopes of a single channel usually are irregular. Depressions develop along the channels and from place to place along their side slopes. All of these are favorable traps for gold. Usually the deepest channel will contain the most gold and its gravels near bedrock will have the highest grade. This is because greater amounts of gravel have been transported along this than along shallower channels. This holds true for coarse gold in single alluvial placer cycles, but cycles interrupted by glaciation may differ according to the vigor of each glacial stage and the length of the intervals between them.

On a larger scale, gold accumulates in two places: in depressions or "holes" along the channel bedrock and at the heads of flat stretches immediately below steeper stretches. The first are traps for coarse gold. They are not common and range from large potholes, which may have rich accumulations, to longer, shallower and poorer scours. The second are often sites for significant coarse gold accumulation on bedrock and minor accumulations of flour gold on skim bars.

Fine Grained Gold

In general a stream's course in plan view (as seen on a map) determines favored sites only for accumulation of fine gold. Float gold is deposited near surface on accretion bars at places where stream velocity decreases; particularly at the upstream points of bars on the inner, concave sides of stream bends. Bedrock gold accumulation at such places is not significant except in the case of streams flowing in narrow valleys in incised meanders. Elsewhere, accumulation of coarse gold on bedrock is controlled largely by bed and bedrock irregularities seen in sectional views.

FAN PLACERS

Fan placers differ from the other types of alluvial placers. They are placers on alluvial fans, deposits of alluvium dumped by streams where they flow out of steep mountain canyons onto more gently sloping lowlands. The streams flow first along one course, then along others, depositing material and forming a cone-shaped deposit. Placers may be formed on the upper part of the fan along several channels, rather than a single one. These channels usually do not reach bedrock; their floors are in fan material. Younger placers lie in material deposited after, and may lie over, older placers, which is the contrary of other alluvial placers. Terrace No. 2 and the outwash plains below the mouth of the canyon of Tarryall Creek in Park County have fan characteristics, but this fan has not been built up high at its head, and the conical shape is not readily apparent. The placer deposits of Buckeye Gulch in Lake County and Blank Gulch/Placer Creek in Chaffee County are on alluvial fans.

LAKES

There are no placers in Colorado's lakes. None of the lakes are large enough to have significant wave action or longshore currents. Consequently, when a stream enters a lake its bedload sediments, together with the placer gold the stream may be moving, are deposited together in a delta deposit. There is gold accumulation but no concentration.

GLACIO-FLUVIAL PLACERS AND EFFECTS OF GLACIATION

Glaciation, alone, does not form placers but rather leads to the destruction of pre-existing ones in its path. Glacial erosion is accomplished by plucking and by abrasion. Glacial plucking, as ice pulls away from a rock face, forms large blocks and does not liberate gold. Glacial abrasion, as rock-laden ice under great pressure grinds over bedrock, produces rock flour. Such grinding of a gold-bearing bedrock deposit would produce very fine gold fragments. There is no placer concentration within or beneath actively moving ice. Instead, as material from a pre-existing placer or glacially eroded gold-bearing material from a bedrock deposit is transported by the glacier, it is mixed with barren debris and quickly diluted.

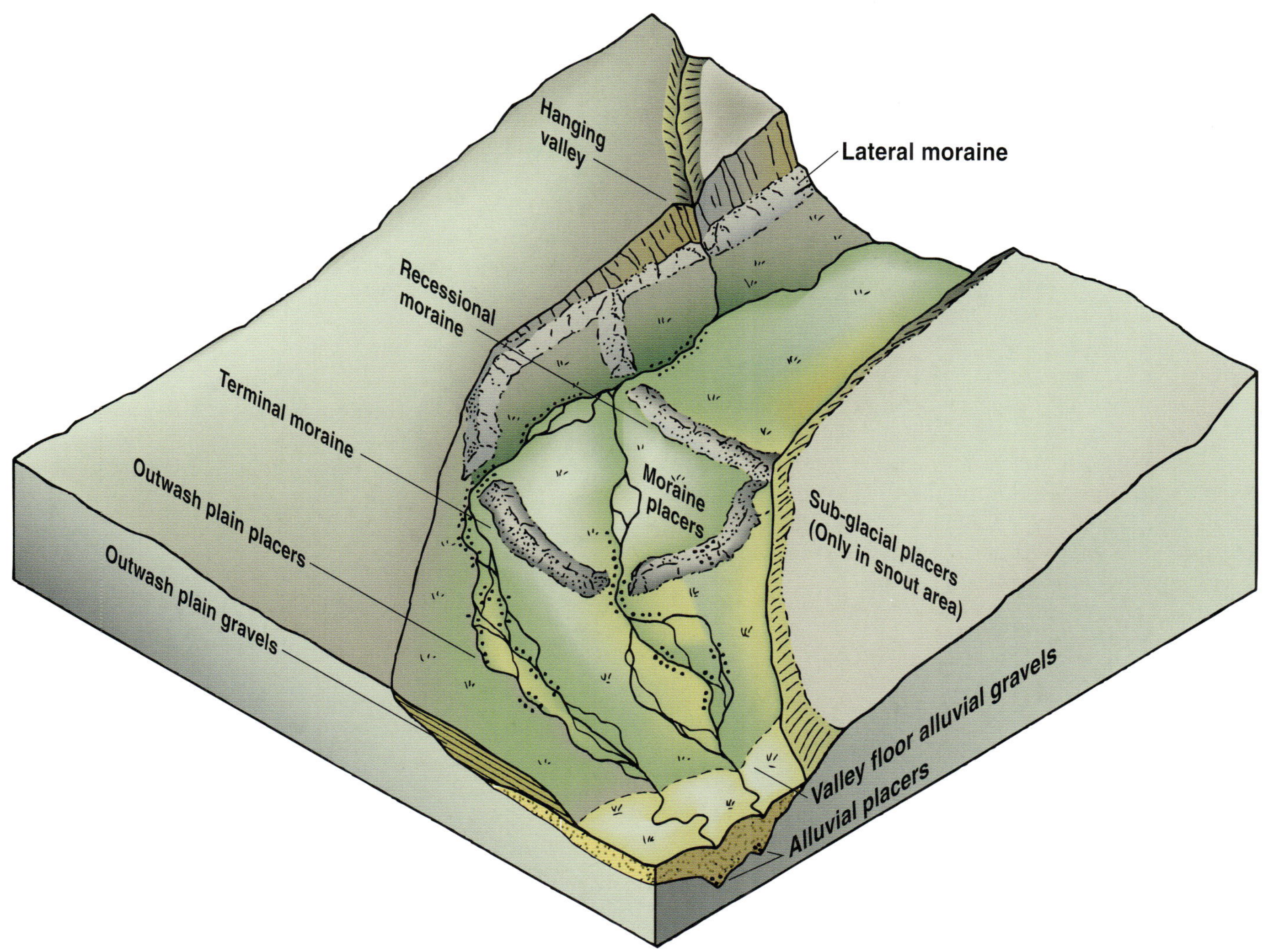

Figure 32. Glacio-fluvial placer types.

However, there may be placer formation in meltwater streams downstream from or beneath the snout of the glacier. Meltwater streams, particularly in times of glacial retreat, are strong and can wash and concentrate large volumes of debris. Placers in whose formation glacial erosion and transportation as well as alluvial processes play a role are called glacio-fluvial placers.

Glacio-fluvial placers have three settings: in sub-glacial deposits, in terminal or near-terminus moraines and in outwash plains. During times of glacial retreat, when the ice is not advancing, if the till beneath a glacier is gold-bearing, meltwater streams can wash and concentrate it, forming sub-glacial placers. These usually are not extensive, and are destroyed if ice movement resumes.

If the terminal moraine of a glacier or the moraines near its snout are gold-bearing, meltwater streams flowing across this material wash it, concentrate and redeposit the coarser gold in channels in the moraine and transport the finer gold onto the outwash plain and perhaps into the stream beyond. If the glacier is in stillstand or retreat and the ice is not moving, the terminal moraine placers are preserved. If the glacier is advancing, the sub-glacial placers may be overridden and/or pushed and the moraine and any placers in it may be pushed ahead, all the while being reworked by meltwater streams.

Coarse gold is concentrated near the moraine. The courses of terminal moraine placer channels are controlled by surface features and not by bedrock; they may go around the ends of the moraine or cross it at any place and may not reach bedrock or may be incised in it. The finer gold from the terminal moraine is swept onto the outwash plain where outwash placers may form.

If there are pre-existing placers, they are reworked and enriched with outwash gold. The pre-existing placers may be completely or only partly reworked. At the South Platte River below Fairplay in Park County outwash gravels deposited by the first, Bull Lake, glaciation rest on bedrock; the outwash gravels of the second, Pinedale, glaciation lie over the first on a clay false bedrock. (False bedrock is a comparatively hard or impermeable layer on which placer gold accumulates within a placer some distance above true bedrock. It shows that the gravel lying above it was deposited by streams less powerful than those that deposited the gravels beneath it that lie on true bedrock.)

Colorado's glaciations greatly affected the placers. In valleys that were not glaciated placers developed without interruption from the time their bedrock sources were first exposed to erosion in the canyon-cutting cycle. (Examples are placers in the Grayback district in Costilla County and in North Clear Creek watershed in Gilpin County.) Placer formation in glaciated valleys was interrupted if ice covered the bedrock source areas. (The absence of significant placers west of the South Platte River between Alma and the heads of Mosquito and Buckskin Creeks in Park County, important sources of gold, shows that the glaciers that covered all of these valleys stripped away the placer deposits that probably were formed here during pre-glacial and inter-glacial times.) The glaciofluvial processes just discussed modified placer formation at the snouts of the glaciers and affected the placers that lay downstream from the glaciers, flooding them with meltwater and glacial debris. (Examples of the first are the Alma and Snowstorm placers in Park County; of the second the Fairplay placer in Park County and the placers downstream from Sawpit on the San Miguel River in San Miguel County.)

EOLIAN PLACERS

An eolian placer is a placer deposit formed by wind action. Wind erosion can slowly sandblast bedrock and liberate gold particles, but these are smaller than those that would be liberated by weathering. Although concentration can be done in air, wind alone is not a very effective placer former. As explained elsewhere the ratio of the effective specific gravities of gold and rock waste is 1½ times greater in water than in air. Air concentration cannot be accomplished if the gravel is not thoroughly dry or if the gravel is cemented. It also requires that the gravel be fairly well sized. Stated another way, the strongest winds are weak compared to streams of water and will move fragments of rock waste only about large pebble size; but they will carry away flour and probably very fine gold. Larger gold would scarcely be transported. Wind does not move or transport its load along channels. So, unless conditions are ideal (that is, the immediate source gravel is less than cobble size, nearly dry, uncemented and contains coarser than flour gold), the best placer wind action can form is a deposit less than cobble diameter deep, more or less coextensive with the immediate source outcrop and of a lower grade than most alluvial placers, unless the grade of the source is phenomenal. Nevertheless, in some desert regions economic eolian placers have been discovered.

In Colorado some dune sand in Saguache and Alamosa Counties, partly in the Great Sand Dunes National Park is gold-bearing, but none of the placers are economic.

MARINE PLACERS

Marine placers are placer deposits formed by marine processes or lying beneath the sea. Marine placers of the first kind, beach placers, are formed by longshore currents and wave action on beaches. A bedrock or immediate source on a shoreline can be eroded and a placer formed by wave action. If a stream delivers gold-bearing gravel to a shore, wave action may concentrate it there or longshore currents may transport it to other near-shore sites where wave action can concentrate it. These beach placers, after formation, may remain at sea level or be either elevated or submerged. Marine placers of the second kind are submerged placers. Originally they may have been beach placers or placers of any other type, and through land subsidence relative to sea level they have been submerged beneath the sea. The placers of Nome, Alaska, are good examples of marine placers; there are none in Colorado.

FOSSIL PLACERS

Fossil placers are placer deposits, originally of any type, that have become buried in the stratigraphic section beneath younger sedimentary rocks and/or lavas. The placers that were worked in the drift mines at about 7,000-feet elevation in the Iron Springs Divide area in Moffat County are fossil placers, originally alluvial and perhaps marine deposits, contained in the Eocene Cathedral Bluffs member of the Wasatch formation. Fossil alluvial placers in the Oligocene Castle Rock formation at Cherry Creek Divide in Douglas Counties were the bedrock source of much of the gold of the Cherry Creek watershed.

GRADES OF PLACER DEPOSITS

The grades of most placer deposits probably will seem surprisingly low to many laymen. Table 3 gives some grades in terms of value of gold per cubic yard and of the ratio by weight of gold to gravel. In the booming discovery days the little data that was published was mainly in terms of dollars per hand per day. The data given in the table for 1862 is clearly exceptional—there were not many pansful of either value. The pans reported were taken from stream placers and probably from near bedrock.

October 16, 1860, the Denver Rocky Mountain News reported that most successful miners at Grass Valley Bar, on the south side of Clear Creek just east of Idaho Springs, were making from $12 to $38 per day, but that yields were beginning to decrease. In 1863 miners were said to be producing $8 to $15 per day at Stilson's Patch, said to be the most profitable area in the Breckenridge district at the time. In 1871 men working in North Clear Creek near Black Hawk were reported to be producing $11.60 per man per day. The Idaho Springs placer was a bench placer, probably worked by shoveling-in, with a productivity of up to 10 cubic yards per man per day. Stilson's Patch was probably a colluvial placer worked in the same manner. The North Clear Creek placers were stream placers being worked with a bedrock flume, an operation similar to shoveling-in, with productivity probably no more than five cubic yards per man per day. So these placer gravels, being worked for the first or second time, must have yielded from $1 to nearly $2 per cubic yard ($1 to $4 or better at the very early Grass Valley Bar operation.)

importantly, because the placer stream is joined downstream by other streams with barren gravels and the placer gravels are diluted. So the source of any colors in a pan is upstream to wherever the highest and most encouraging samples are found and from there upslope.

The prospector took advantage of this in his searches. Finding a few colors in gravels he would move upstream panning samples every mile or so, watching his recovery. Ideally, over a few miles the gold colors would increase in number and in size. At stream junctions he would take samples on each fork and follow the better sample or perhaps the one of the two with colors. As the pan tails became longer and the colors larger, he reduced the distance between samples until he found a place where the tails became notably poorer or even disappeared. At this point he would turn to the slopes, figuring that the bedrock source would be upslope.

On the slopes he would pan soils or dig into the colluviums for samples and follow the colors upslope. Frequently, since he was looking for a bedrock source, he pulverized rock fragments in his samples and panned their powders. He especially selected fragments of quartz, of altered rocks and of yellow or rusty red, limonite-stained rocks. He dug pits or trenches to see bedrock, to look for a gold deposit beneath the cover. Sometimes he found nothing, sometimes he found terrace gravels and other times he found a bedrock source of gold.

Table 3. Placer grades.

Placer	Kind of Mine	Year	Dollars per pan	Dollars per cu yd	Ounces per cu yd	1 part in
Nigger Gulch	Shoveling in	1862	3.64	655.20	41.08	1,065
N. Clear Creek	Shoveling in	1862	0.75	135.00	8.47	5,165
Alma	Drift	1937		3.00	0.1038	421,304
				18.00	0.6231	70,217
Idaho Springs	Drift	1934		17.86	0.55	79,544
Fairplay	Hydraulic	1869–1872		0.43	0.0269	1,622,575
Clear Creek	Dry land plant	1935		0.61	0.0207	2,112,152
Alma	Dry land plant	1937		0.41	0.0142	3,082,152
				3.60	0.1246	351,134
Derry Ranch	Dry land drag line plant	1947–1950		0.19	0.0072	6,094,258
Breckenridge	Tonopah dredges	1914–1925		0.12	0.0074	5,891,169
Fairplay	South Platte No. 1	1841–1952	Dredge	0.09	0.0031	14,016,275

Within a few years—two or three at many places in the Central City and Breckenridge districts—the grades of the gravels being worked diminished. After this, new methods were intermittently developed which increased productivity and permitted working previously inaccessible and/or subeconomic gravels. Table 3 shows these changes.

BEDROCK PROSPECTING

You have seen that gold transportation is controlled by gravity—gold moves very, very slowly downhill and downstream. During transportation gold fragments become smaller and more beaten. Although there will be places in streams where gold concentration is favored and others were not, overall, gold concentration will decrease downstream. This is because gold accumulation becomes less efficient as colors become smaller and, more

WHERE TO PAN

RESPECT PRIVATE PROPERTY

Much of, but not all, Colorado's placer ground is private property. Just as when fishing or hunting, before panning on fenced or posted land, get permission! Do your shoveling and dispose of your tailings so they are inconspicuous, if not invisible! Before using a rocker or a portable dredge, get the land-owner's permission or if in a national forest a Ranger's.

DANGERS

Although panning is not dangerous, there are a few special risks that you should avoid:

• Don't go into drift mines or other abandoned workings! All the drift mines have been abandoned at least 60 years and any abandoned working is hazardous. A mine needs continual maintenance to be safe.

• Be especially careful around streams more than knee-deep. The streams are deceptively swift, powerful and cold. Rocks on the stream beds often are moss-covered and very slippery.

• Be careful around water-filled dredge ponds or other placer pits. They are often deep with sheer banks and the water is cold. Avoid the banks. If you must be near them, locate a place you could climb out should they collapse.

• Do not do scuba placering. Compared with the streams elsewhere where it has been done with more or less success, Colorado's mountain streams are too small and too steep for safety. Also, many of these streams' attractive sections have already been cleaned to bedrock one or more times.

MAPS

The U. S. Geological Survey publishes the County Map Series at 1:50,000 scale, 30 X 60 Minute Series maps at 1:100,000 scale (metric), and 7.5 Minute Quadrangle Series maps at 1:24,000 scale. All these maps show topography and name gulches, streams and other local features. They are for sale at the U. S. Geologic Survey's Map Distribution Office in the Denver Federal Center; Building

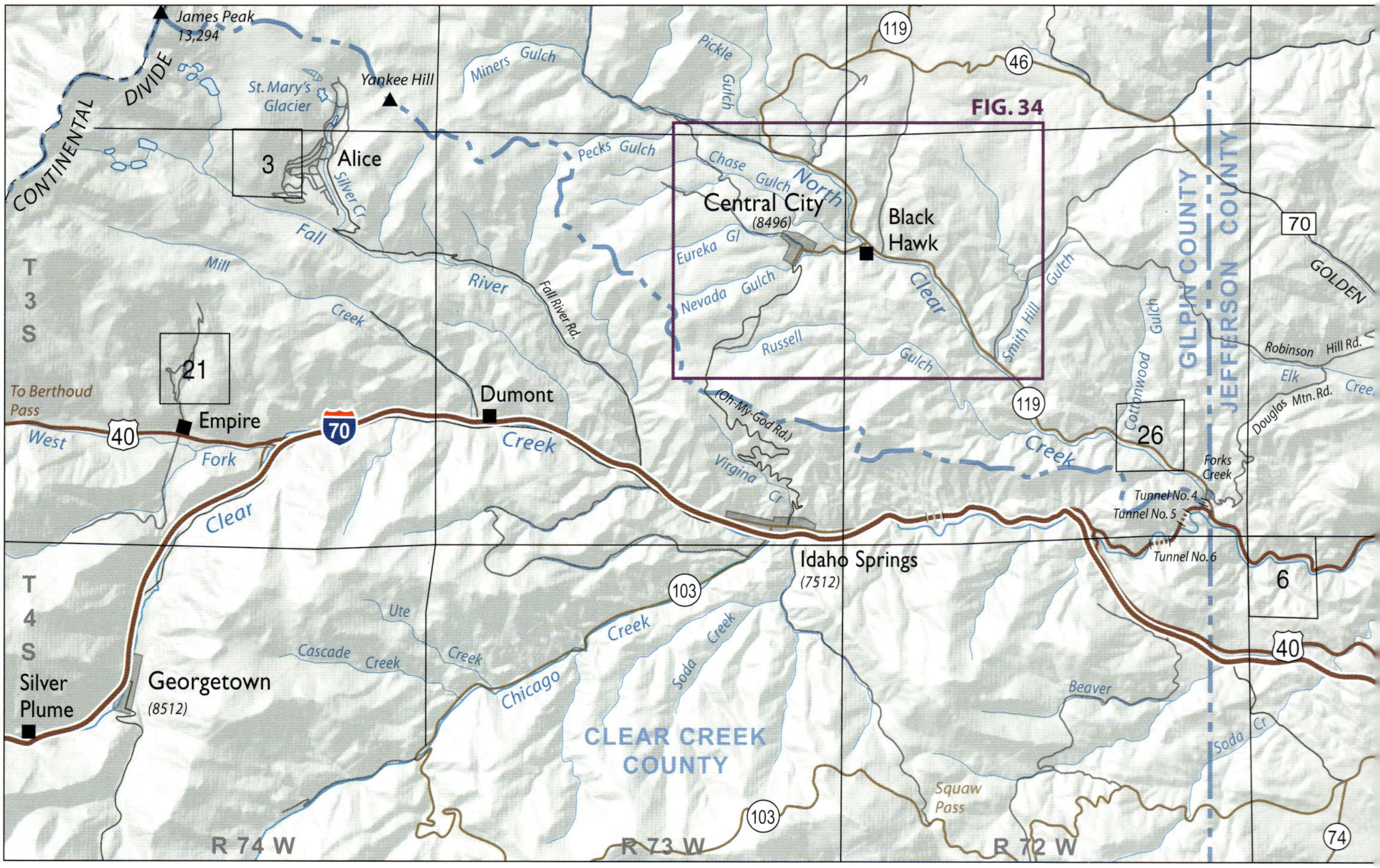

Figure 33. Location map of Clear Creek and tributaries placers, Adams, Clear Creek, Gilpin, and Jefferson Counties.

810, P. O. Lakewood, Colo., 80226 (303-236-5900), as well as some local stores.

The U. S. Forest Service publishes maps of the National Forests and surrounding areas that do not have topographic detail as the U.S.G.S. maps, but are usually adequate, nevertheless. You can also preview these maps online at *www.fs.fed.us/r2*. These maps are also for sale at the U.S.G.S. map store at the Federal Center and at Ranger Stations throughout the State.

GOOD AREAS FOR PANNING

The following placers are ones where you are likely to recover colors washing a few pans of gravel. They are also easy to locate.

ADAMS, CLEAR CREEK, GILPIN AND JEFFERSON COUNTIES

Clear Creek and Tributaries

Placers have been worked from place to place from Jackson's Bar on Chicago Creek, from the mouth of Fall River on Clear Creek, from the heads of Russell and Gregory Gulches west of North Clear Creek and from above timberline on North Clear Creek to Denver. This placering began the winter of 1859 and has continued with interruptions to the present. Jackson's Bar and Gregory Gulch were two of the three economic gold discoveries early in 1859 that sustained the Colorado gold rush. No commercial placer mining continues today.

Going up North Clear Creek from Forks Creek (at the junction of U.S. 6 and Colorado 119) there are signs of placer mining along the creek almost continuously to Pickle Gulch and in Russell and Gregory Gulches and many of their tributaries nearly to their heads. All have been placered from once to more than three times. The valley placers were worked by shoveling-in; where water was available (especially in the upper courses), by small scale hydraulicking, and along North Clear Creek from Black Hawk downstream by drift mining and by dragline and mobile dry land washing plants. Along North Clear Creek below Black Hawk you can see a few placer sites from place to place. A prominent one is in sec. 26, T. 3 S., R. 72 W., on gravel terraces about 55 feet above the creek. There were eluvial placers along many veins in the district, and between them and the streams colluvial placers were worked in some places.

You can recover some colors fairly quickly at any of these sites, Along the main creek you cannot reach bedrock; in the channels there the valley gravels are 25 to 30 feet thick. Look for skim bars and for gravel on the downstream sides of large boulders. Of the gulches, Russell Gulch watershed and Gregory, Nevada, Bobtail and Packard Gulches are most promising but they are ob-

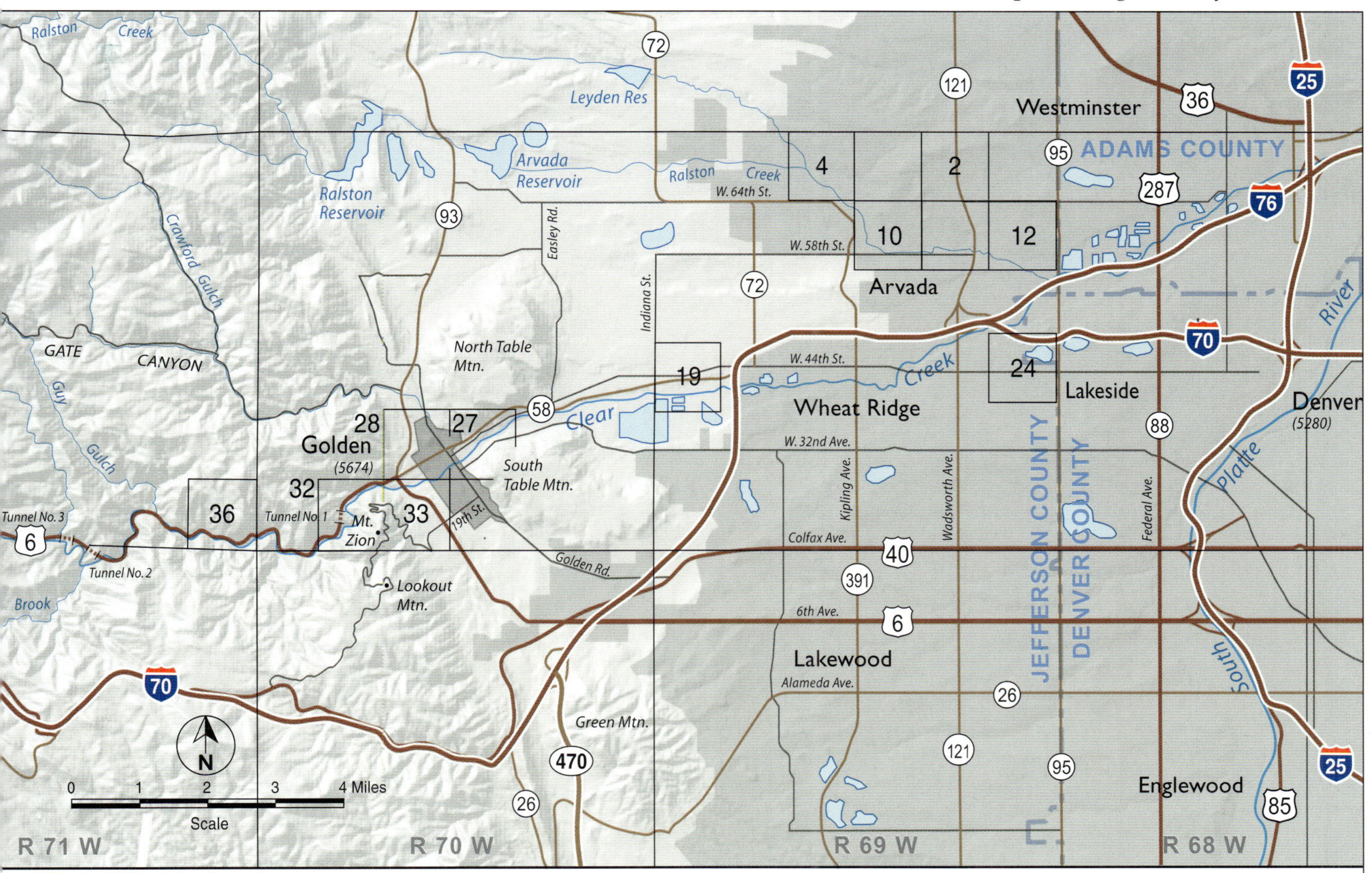

structed in places, especially with the recent developments in Central City and Blackhawk. On the terraces you can reach bedrock gravels in many places. The eluvial deposits are almost completely exhausted, but you might get a panful or two of good dirt from them. A good place to start would be climbing along the Gregory vein, marked by a line of pits and depressions that go uphill southwest from the Gregory monument beside the road between Black Hawk and Central City. You will see what a vein outcrop looks like and can then recognize others leading from some of the shafts in the district. In places along the outcrops underground workings are open. Do not enter them! Most have not been worked for 80 years or more and are unsafe. The backs (roofs) and in many places even the floors are not secure.

Placers have been found from place to place from the mouth of Fall River downstream to Forks Creek and along Chicago Creek for about half a mile upstream from Clear Creek. The Chicago Creek section was "Jackson's Bar". The site of Jackson's discovery was on the west side of Chicago Creek about a quarter of a mile south of Clear Creek. Between Fall River and the west end of Idaho Springs was "Spanish Bar". Clear Creek heads in higher mountains than does North Clear Creek, in areas that underwent glaciation; whereas North Clear Creek's watershed did not suffer glaciation. The principal gold deposits lie in valleys that join Clear Creek below Fall River. (The gold deposits of North Empire (page 62) and of Alice (page 61) each gave rise to eluvial and colluvial placer deposits in tributary valleys, but glaciers and heavy outwash streams destroyed or buried any stream placers derived from them in the Clear Creek and Fall River valleys.) Consequently, there are no important placers on Clear Creek above Spanish Bar. Also because of the volume of the outwash streams, the valley placers from Spanish Bar downstream have thicker gravels and more and bigger boulders than do the placers on North Clear Creek.

There are remnants of terraces 25, 55, and 180 feet above creek level near Idaho Springs and at a few places downstream to Forks Creek. Some placer gravel can be found at some of them and on the point bars and below boulders on Clear Creek.

From Forks Creek to the mouth of Clear Creek Canyon gravel piles beside the stream bed mark the site of dry land washing plant placer operations of the early 1930s. The banks beside some of them expose gravel banks beneath hillside mantle. There are a few terrace remnants in Clear Creek Canyon in sec. 6, T. 4 S., R. 71 W., and in sec. 36, T. 3 S., R. 71 W., that correspond to the 55-foot terrace in Idaho Springs and North Clear Creek.

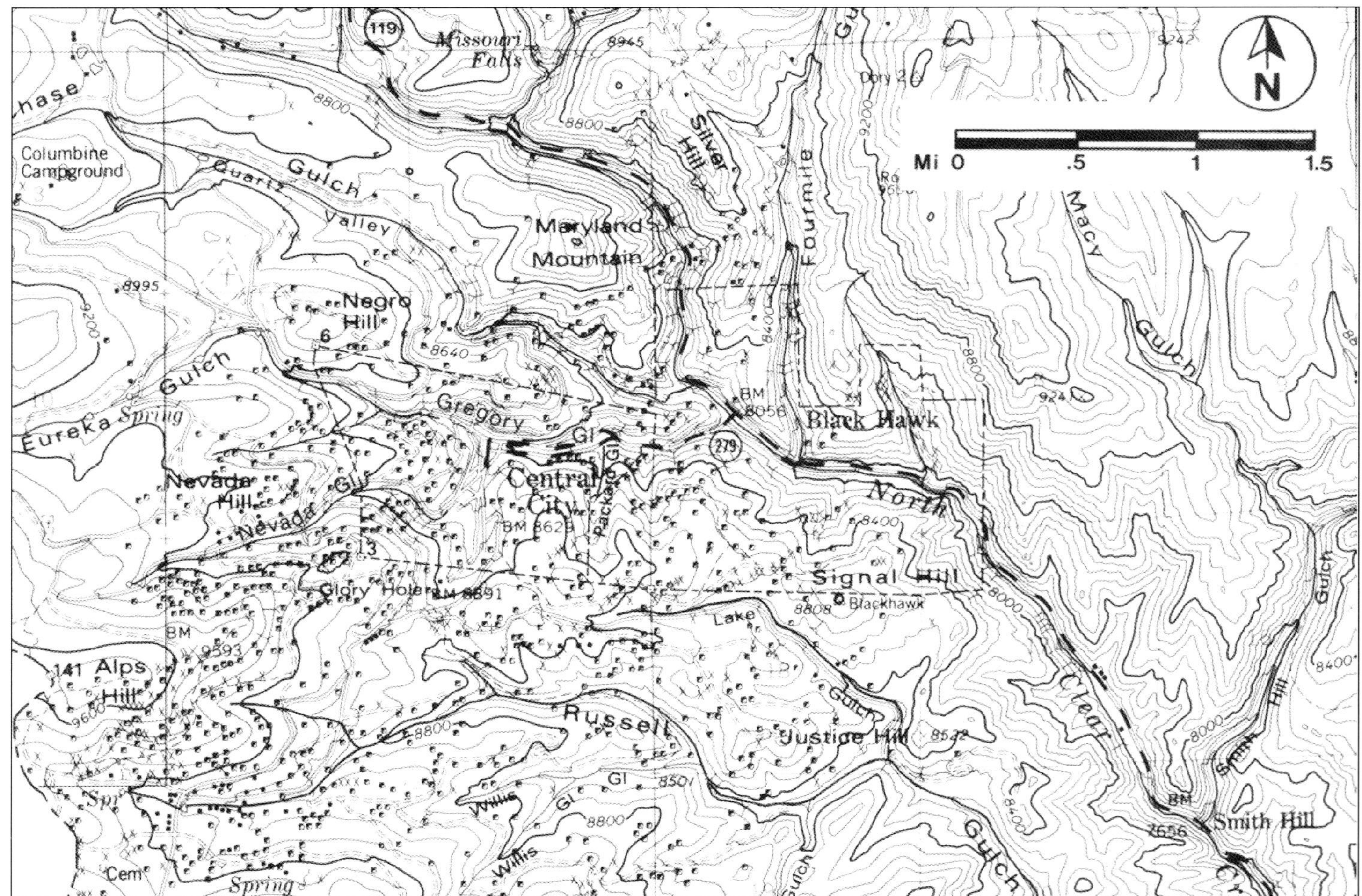

Figure 34. Topographic map of Central City and Black Hawk area, Gilpin County, (U.S.G.S., County Map Series).

These were worked at least in part in the 1930s. On the north slopes of Mount Zion (the mountain with Colorado Mines' "M") west of Golden, in secs. 32 and 33, T. 3 S., R. 70 W. there are old cuts in a terrace about 230 feet above the creek. These expose some good grade gravels. They probably weren't economic—it was as hard for the miners to bring gravel to the creek as it will be for you if you attempt it.

From the mouth of the canyon downstream to the South Platte River, the gravels beneath Clear Creek valley are gold-bearing. In 1904 and 1905 attempts were made to work them with two dredges located in NE sec. 27, T. 3 S., R. 70 W. This operation failed because there were many lava boulders too big to move on bedrock. Placer miners worked at the mouth of the canyon in 1892 and 1893 and in the early Depression years. Those working in the 1890s were reported to be making \$3 to \$5 per day, which would correspond to a production of one-sixth to one-quarter ounce per day. The Depression miners rarely made this much.

These miners did some drifting and also worked the skim bars. At the mouth of the canyon Clear Creek's gradient becomes more gentle and some gold can often be gotten from the top gravels at the heads of the point bars through Golden after the spring runoff and after heavy storms.

The sites in the canyon and the point bars in Golden are favorable sources of gravel for panning.

PARK COUNTY

Fairplay District

This district in Park County has many placers. Although shows of gold have been found from the foot of Hoosier Pass downstream and along Buckskin and Mosquito Creeks to Alma, the principal placers are at Alma, along the South Platte River from Alma to below Fairplay, along lower Beaver Creek and between Beaver Creek and the South Platte about midway between Fairplay and Alma (Snowstorm or Timberline placer). An unusual placer at the edge of the district lies on Pennsylvania Mountain. The Pennsylvania Mountain placer is a colluvial placer; the others are glacio-fluvial placers.

Although the principal bedrock gold deposits of the area lie to the west of the South Platte River, the flow of the Wisconsin glaciers was such that the ice transported their gold to the central and eastern parts of the South Platte valley. The gold of Beaver Creek upstream from sec. 20, T. 9 S., R. 77 W. came from the Mount Silverheels area. Most of the gold of lower Beaver Creek and all the gold of the other Fairplay district placers, except Pennsylvania Mountain, came from the western part of the South Platte watershed—Mosquito and Buckskin Gulches and North Star Mountain. The source of the gold of Pennsylvania Mountain has not been identified.

From the earlier Illinoian glaciation (Bull Lake) there remains only the breached terminal moraine at Fairplay, some remnants of lateral moraine and extensive outwash buried beneath the outwash plain south east of Fairplay. The Wisconsin glaciation (Pinedale) is marked by terminal moraine beside and upstream from the earlier Bull Lake one, two recessional moraines between Fairplay and Alma and the upper, exposed layer of outwash on the plain southeast of Fairplay. (A recessional moraine is a moraine that is formed upstream from the lowest, or terminal, one during a minor surge of glacial activity.)

The Alma placer lies in a hill to the east and southeast of the town of Alma on the east side of the river. The placer is composed of alluvial deposits formed between the last two glacial surges overlain and mixed with gold-bearing till from the last surge's lateral moraine in which some glacio-fluvial meltwater channels formed. Some of the gravels from the bottoms of the channels are extremely rich, but they are also extremely patchy, as the description suggests. Here, pick gravel for panning from the sections with most cobbles, and be careful around the cut faces; from time to time gravel falls!

The next important placers downstream are those in the two moraines and their outwash from Sacramento Creek through the town of Fairplay. The upstream parts and those in the moraine were worked by hydraulicking from 1872 intermittently through the 1890s. The downstream part near river level was worked by the first South Platte dredge in 1922-1925. This deposit includes gold-bearing till and meltwater channels in the moraines and outwash deposits in the valley cut through them. The banks close to the river and the river, itself, should provide the best panning results.

South of Fairplay on the outwash plain on the west side of the River extensive boulder piles mark the area mined by the South Platte No. 1 dredge. This plain is underlain by outwash gravels of the early, Bull Lake glaciation lying on bedrock and overlain by later, Pinedale outwash gravels lying on a clay false bedrock. Although the surface gravels carry some gold, the bulk of the gold lies on bedrock and false bedrock, out of reach. The best gravel here is probably along the river, where the outwash gravels have been eroded and slightly reconcentrated.

On the slopes east of Colorado 9 in sec. 19, T. 9 S., R. 77 W., is the Snowstorm or Timberline placer and in secs. 20, 29 and 30 is the MacConnell cut. Placers here were worked in the 1870s, the Snowstorm was hydraulicked in 1902 and 1903 and the MacConnell cut worked from 1904 to 1916. The Timberline dragline floating washing plant was installed in 1941 and worked in 1942, 1946, and 1947. These placers are deposits in meltwater channels in Bull Lake and Pinedale Wisconsin lateral moraines. These moraines contain masses of alluvial placer gravels formed in interglacial times, mixed with till and pushed onto the hillside by the glaciers. Gold was concentrated on bedrock, but in the southern part of the placer there were two false bedrock streaks, as well. For panning select the lowest and coarsest gravels to be found in the old workings.

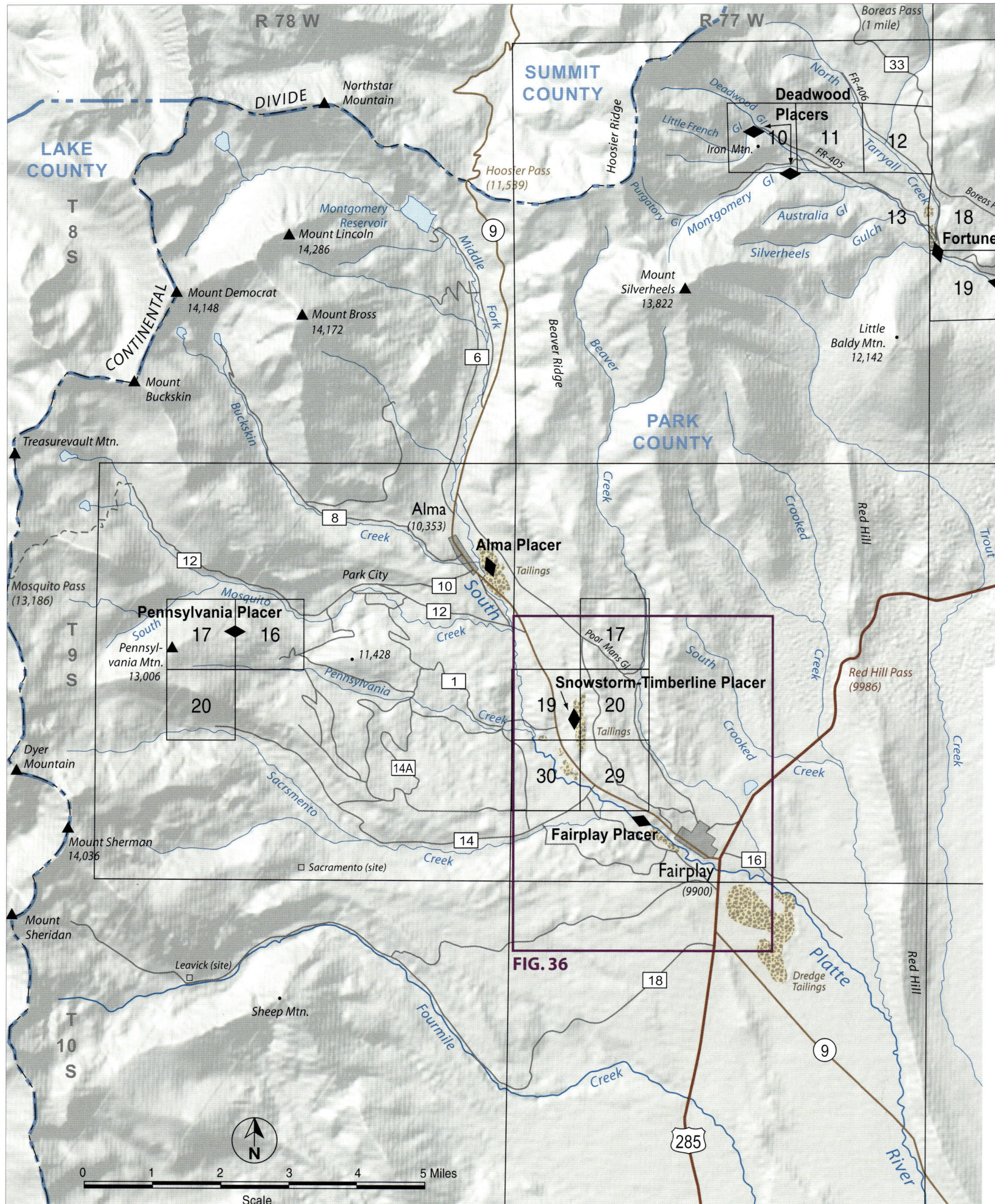

Figure 35. Location map of Fairplay-Tarryall district placers, Park County,

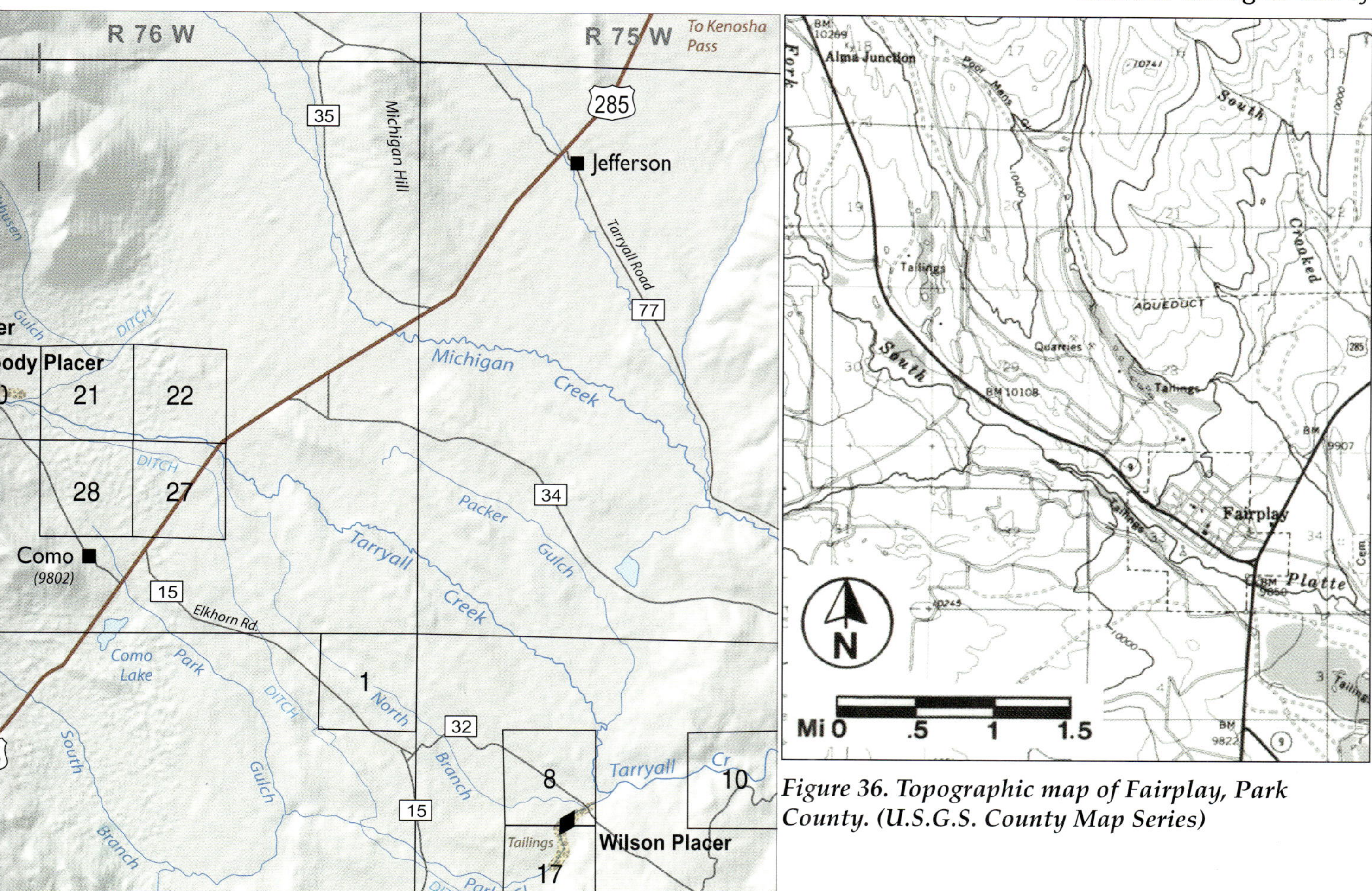

Figure 36. Topographic map of Fairplay, Park County. (U.S.G.S. County Map Series)

Along Beaver Creek upstream to the southeast corner of sec. 17 and from there to the northwest in Poorman Gulch for a short distance are placers formed chiefly by meltwater streams flowing along the edge of Pinedale moraines. These placers were of good grade, profitably worked. Upstream from the southeast corner of sec. 17 there are distinctly poorer stream placers to the slopes of Mount Silverheels.

Pennsylvania Mountain

The Pennsylvania Mountain placer is on the upper slopes of Pennsylvania Mountain in secs. 16, 17 and 20, T. 9 S., R. 78 W. at an elevation of about 12,250 feet. It lies between Pennsylvania Gulch and the cliffs overlooking Mosquito Creek on an unglaciated slope that has a bedrock trough extending from near the center of sec. 17 to near the center of sec. 16. Most of the old placer workings are along this trough. On the slope there is up to 19 feet of mantle material composed of angular quartzite and rounded porphyry cobbles and boulders in a sand matrix.

The gold is somewhat flattened but still has sharp edges and often shows crystal faces. The placer is well known for the coarse gold that has been produced there. The largest nugget produced weighed 11.95 ounces—"a pennyweight less than a pound." In 1990 Shane Dodge found an 8.6-ounce nugget and a 12-ounce one was found in 1937. There have been many one-ounce samples. The gold produced is notable for the low proportion of fine gold that it contains. The gold is not concentrated on bedrock but rather in the central part of the gravel section. This is a colluvial placer. The bedrock source of the gold has not been identified.

A great problem at this placer has been the scarcity of water. At one time the gravel was carried by aerial tramway from the placer to Mosquito Creek. Other times water has been caught in dams on Pennsylvania Creek and recirculated. Neither solution was very effective. Should you prospect here, look for gravel about midway between the gulch on the Mountain and the cliffs in an area where there are many old pits. Remember that with such coarse gold, good grade gravel will contain only a few fragments in a cubic yard; there will be many barren pans. The placer area is private property—***get permission before you go there!***

Tarryall Creek Area

Gold was discovered near the forks of Tarryall Creek in late summer of 1859 and worked, usually on a small scale, with some interruptions until the 1890s. Many individuals placered here during the Depression. A dry land plant was set up on the Wilson placer in 1934 and worked until 1942 and in 1941 to 1942 and 1946 to 1947 a dragline dredge was operated on Tarryall Creek below Peabody's and on Cline Bench. At about the same time there was extensive work in Little French Gulch. In the 1980s there was again some activity in the valley.

The important placers can be considered in three groups: 1) Those in the canyon of Tarryall Creek (For tune and Peabody placers), Middle Tarryall Creek, Little French Gulch, the lower mile of North Tarryall Creek, and smaller placers in Deadwood and Montgomery Gulches. 2) Those on Tarryall Creek in South Park. The valley floor of Tarryall Creek from the mouth of the canyon downstream for about a mile in secs. 19 and 20 and a bench placer in secs, 21, 22 and 27, T. 8 S., R. 76 W. (Cline Bench placer). 3) The placers on North Park Gulch in sec. 1, T. 9 S., R. 76 W., in Park Gulch in secs. 8 and 17, T. 9 S., R. 75 W. (Wilson placer) and downstream on Tarryall Creek in sec. 10.

The bedrock source of these placers is myriad tiny gold veins surrounding an igneous stock on the north slope of Mount Silver Heels and extending from Montgomery to Deadwood Gulches. Although none of these veins have proved economic to mine, in the aggregate they have provided much gold for placer concentration.

In the extreme heads of Montgomery, Little French and Deadwood Gulches are colluvial placers that in a few spots have been of sufficient grade for limited placering. These are unusual because they have developed since the final disappearance of the glaciers; before then they were beneath the ice. They grade downslope into glacio-fluvial and glacial placers.

The Wisconsin Pinedale terminal moraine is at the narrows in the valley in sec. 18, T. 8 S., R. 76 W., and sec. 13, T. 8 S., R. 77 W. Above this terminal moraine there are recessional moraines about at the mid-point of Middle Tarryall Creek, at the mouths of Montgomery and Little French Gulches and further upstream in Montgomery and Deadwood Gulches. The terminal moraine has been worked extensively in lower Tarryall and North Tarryall Creeks and the slope between them at the Fortune placer. The workings develop meltwater channels across and along the terminal moraine, outwash from the recessional moraine upstream and apparently some glacial till unwashed by later streams. This moraine is unusual in that it is all slightly gold-bearing and some unwashed spots have been found that paid to placer. It is also interesting that the gold in North Tarryall Creek came from Middle Tarryall Creek. In secs. 11, 12, and 13 the Middle Tarryall Creek Pinedale glacier flowed northeastward over the divide between the two valleys. There is a remnant of Illinoian Bull Lake moraine below the narrows to the edge of South Park in sec. 19 and into 20, T. 8 S., R. 76 W. In sec. 19 where Tarryall Creek flows on the north edge of the Bull Lake moraine, all the valley floor has been placered (Peabody placer). These deposits were outwash from the Pinedale glacier and reworked Bull Lake till.

From the mouth of the canyon eastward there are terraces at three levels above the floor of Tarryall Creek. The two higher terraces lie on the south side of Tarryall Creek in secs. 27 and 28 and the lower is on the north side of the creek there. The higher terrace (called Terrace No. 2) was formed by Tarryall Creek in pre-Bull Lake time. The terrace below it is the Bull Lake outwash plain. The bottom terrace is the Pinedale outwash plain, into which Tarryall Creek has cut its modern valley floor. Each of these terraces is more or less gold-bearing for a few miles away from the canyon.

The valley floor of Tarryall Creek, worked for more than a mile below Peabody's, was a Pinedale glacio-fluvial deposit enriched by post-glacial concentration. The Cline Bench deposit was a Bull Lake glacio-fluvial deposit in a bedrock stream channel in outwash. There may be other channels like these beneath each terrace.

The immediate source of the placers on North Park Gulch and in Park Gulch (Wilson placer) is the gravels of Terrace No. 2 and the bedrock source is the veins at the head of Middle Tarryall Creek. The placers downstream on lower Tarryall Creek probably have some Bull Lake and Pinedale outwash gold, as well.

In the stream placers, select gravel for panning from as low in the faces as possible. In Cline Bench and adjoining it on the south side of Tarryall Creek, bedrock may be exposed from place to place. In the placers in the canyon, very coarse or cobbley sections in the cuts may be meltwater channel floors and may be well worth panning. At the Fortune placer, since the till is somewhat gold-bearing, sample the moraine banks from place to place, as well. At the heads of the valleys old pits may suggest the location of better grade spots. The gravels contain lots of magnetite in black sand grains and pebbles and cobbles and abundant epidote and garnet.

SUMMIT COUNTY

Breckenridge District

The Breckenridge district, in Summit County, has been the most productive placer district in the State. It includes many different placers, which will be discussed by type. There are colluvial placers; they grade into alluvial placers in gulches; there are moraine placers in French Gulch, terrace placers and valley placers. This district is the only one in the State in which there was placer mining each year from its discovery in 1859 into the 1960s (with the exception of during World War II). The activity ranged from years in which there were only a few small operations to the discovery years when several thousand miners were at work or to the peak of the dredging period when three dredges were in operation along with other placer activities.

In the district are the Blue River, which flows northward, and two of its principal tributaries from the east, the Swan River on the north and French Gulch to the south.

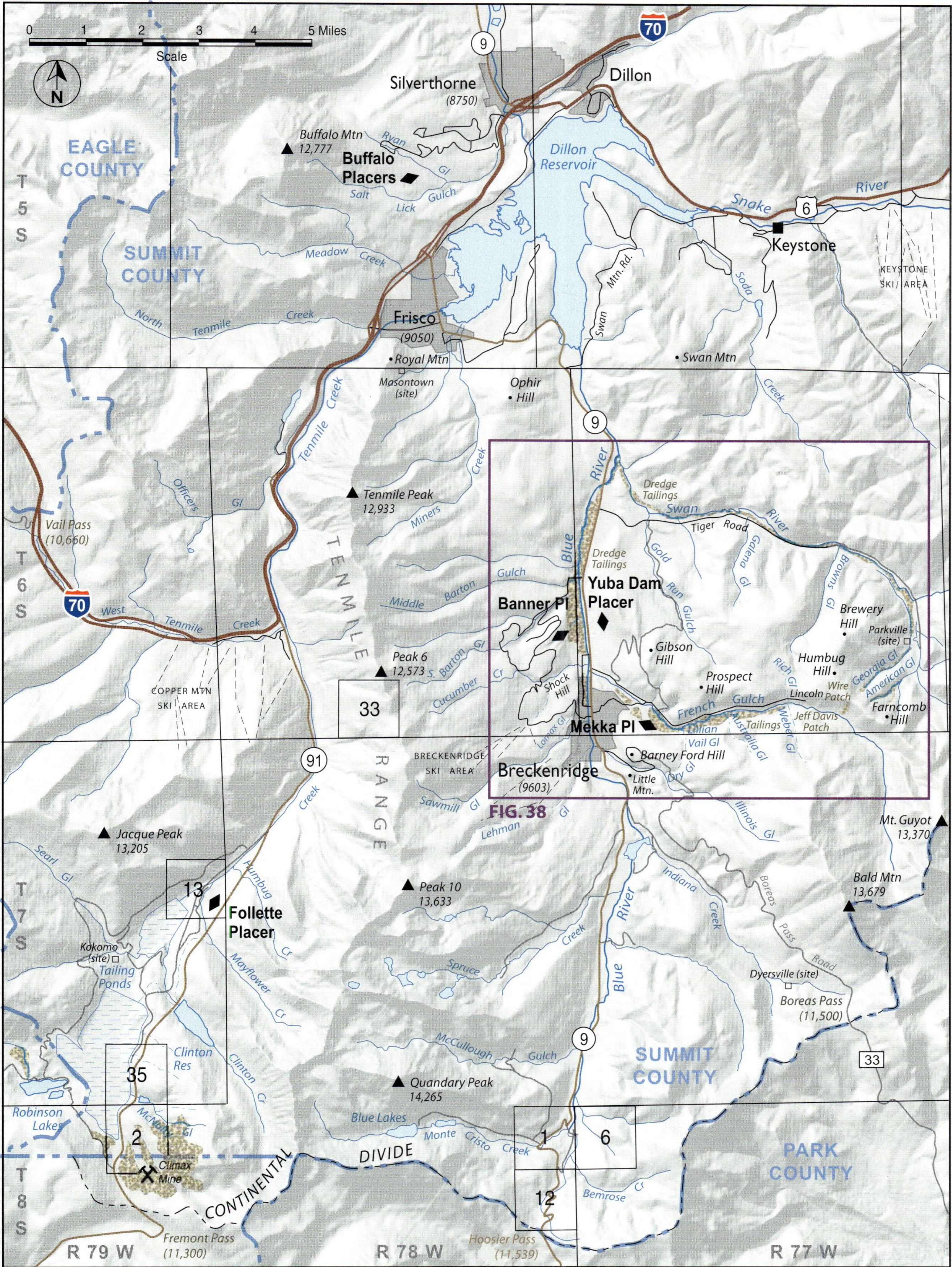

Figure 37. Location map of Breckenridge district placers, Summit County,

Breckenridge is just south of the junction of the Blue and French Gulch. To the west of the district is the Tenmile Range—12,000 to nearly 14,000-foot peaks approximately three miles from the Blue River. Between the Swan and French Gulch is an upland rising to rounded "hills", the eastmost of which, about five miles from Breckenridge, is Farncomb Hill. At the south end of Breckenridge, Illinois Gulch flows into the Blue from the east. Between French and Illinois Gulches is Barney Ford Hill (originally called Nigger Hill, the name used in contemporary accounts and in histories through the 1960s), and just south of Illinois Gulch is Little Mountain. From Barney Ford Hill, on either side of the Blue from place to place to its junction with the Swan and beyond, are gravel-covered terraces usually 200 to 250 feet above the River. Small gulches tributary to these main streams drain the "hills" and the terraces.

Among these, Georgia and American Gulches and the Wire Patch on Farncomb Hill were noteworthy for their exceptionally rich placers. Georgia Gulch shares with upper California Gulch in Lake County the reputation of being the richest and most productive placers (for their size) in the State.

The high peaks of the Tenmile Range are relics of the pre-Summit surface mountains. The summits of Farncomb Hill and the other "hills" between the Swan and the Blue are remnants of the sub-Summit group of surfaces. The high terraces are older than the Illinoian Bull Lake glaciation. Terminal moraines of the upper Blue River glaciers just reach Barney Ford Hill, those of the French Gulch glaciers are about three miles east of Breckenridge, and those of the upper Swan River glaciers extend to the mouths of Georgia and American Gulches. Except in upper French Gulch (at Jeff Davis Patch and Rich and Weber Gulches) and the extreme lower parts of Georgia and American Gulches, the placered areas either were not glaciated or were downstream from the terminal moraines. The gravels worked by the dredges beneath the valley floors were outwash somewhat reworked by post-glacial streams. The gold in the valley gravels was concentrated close to bedrock; there were no extensive false bedrock levels. Apparently here, in contrast to the Fairplay placer, the Bull Lake outwash was largely reworked by the Pinedale outwash streams.

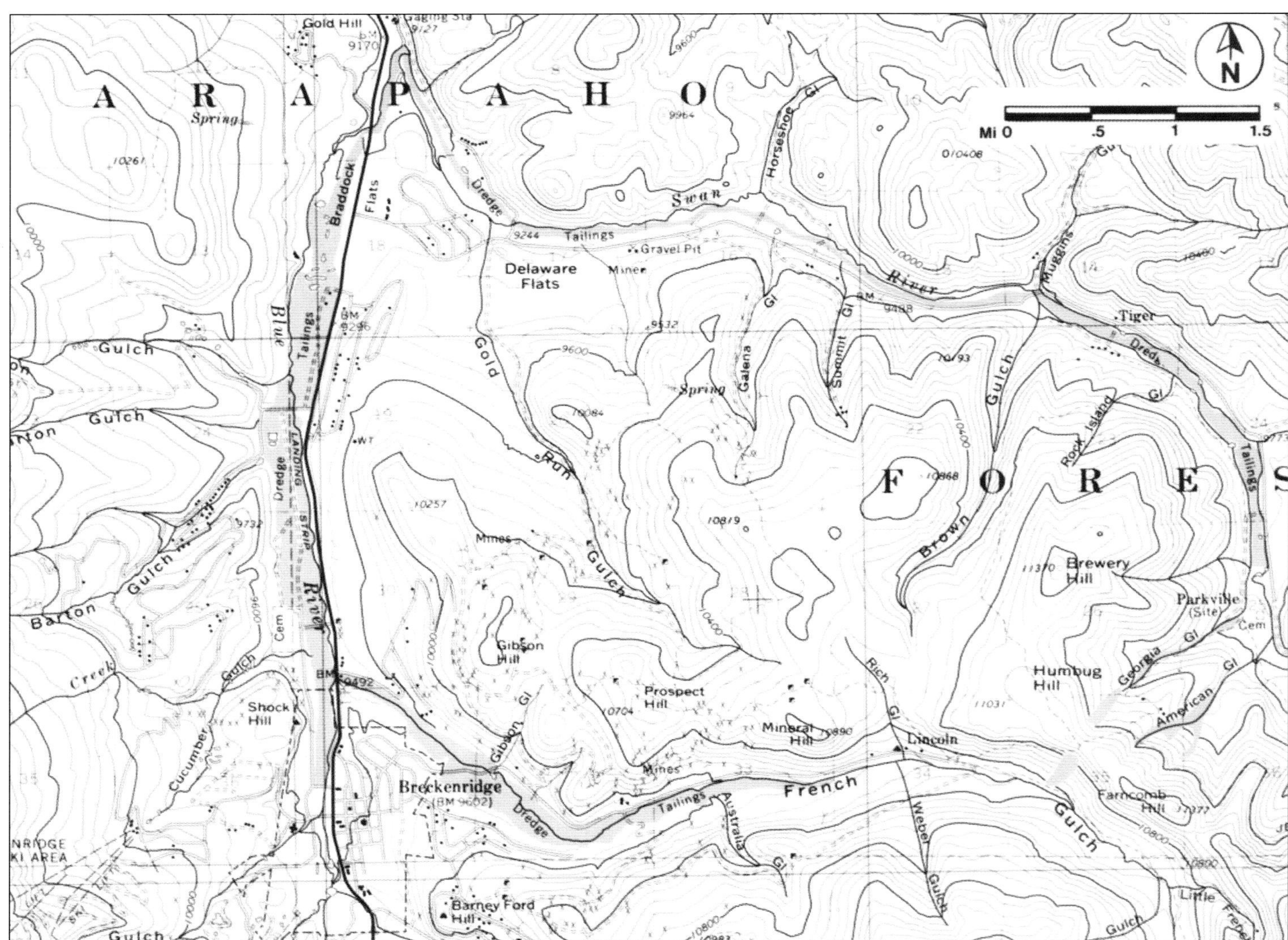

Figure 38. Topographic map of the Breckenridge area, Summit County, (U.S.G.S., County Map Series).

There were numerous sources and a variety of kinds of sources of the placer gold. The source of the gold of Georgia and American Gulches and of the Wire Patch (and a part of the gold in the terraces and major streams) were the veins of Farncomb and nearby Humbug and Brewery Hills. These were many tiny veins—extending hundreds of feet, typically one-half inch thick and nearly barren. They were much explored. Miners might drift on a vein through long distances of barren ground and, if lucky, find a spot an inch wide and two or three feet in diameter filled with hackly wire gold. One day would repay a year's work. There are beautiful specimens of this gold on display at the Denver Museum of Nature and Science and at the Colorado School of Mines library in Golden.

Figure 39. An historical photo looking northerly across Illinois Gulch toward Mayo Gulch, Breckenridge district, Summit County. The floor and lower walls of Mayo Gulch were placered—probably both boomed and hydraulicked. A fan of tailings was deposited in Illinois Gulch in which a monitor is cutting gravels in the foreground. (U.S.G.S, Photo Library, T. S. Lovering, 240)

Veins of this kind, since eroded away, and small stockworks— intensely fractured zones whose many cracks are mineralized—are thought to have furnished the gold of the many hillslope (colluvial) and gulch placers from Galena Gulch to Gold Run and to the Swan River below them. Veins on Barney Ford Hill and Little Mountain supplied the gold to the rich Lillian Vale and Dry Gulch placers in gulches on the east end of Barney Ford Hill and to the terrace and stream placers downstream. Veins in the Dakota Sandstone in Shock Hill, just west of Breckenridge, and in Little Mountain contributed gold to the terrace and stream placers. Finally, some gold must have been contributed to the terrace and stream placers from deposits in the high mountains at the heads of the Blue River and Illinois Gulch.

Nearly all the deposits except those of the last type lay within 300 or 400 feet of the sub-Summit surface. The importance of these deposits as sources is related to their being close to this surface and to the long period of weathering to which the surface was subjected. Gold in the original deposits slowly dissolved in groundwater, moved downward to the water table and reprecipitated in coarser masses that were eventually liberated in the canyon cutting cycle. Gold deposited in this manner is called supergene gold. It is thought that the placers of the area between Gold Run and Galena Gulch lie below the level of the supergene zone in Farncomb Hill-type veins now destroyed—veins that probably supplied the placers' gold.

There were extensive colluvial placers beside and between Galena Gulch and Gold Run, at the heads of Georgia and American Gulches and at the Wire Patch. These deposits, up to 40 feet thick are of angular porphyry cobbles and boulders in a clayey matrix, Those in Georgia and American Gulches and the Wire Patch are nearly worked out. There are still some colluvial gravels on the west side of Gold Run and between it and Galena Gulch. The latter are probably better grade.

The gulch alluvial placer material is very similar to that of the colluvial placers—angular cobbles and boulders in a more or less clayey matrix. Composition of the coarse material varies with the bedrock in the gulch—porphyry or quartzite or both. These placers are confined to the gulches. Nearly every gulch flowing off the "hills" or the terraces has been placered to some degree. Any of the gulches may yield good panning gravel. Gravels as close to bedrock as possible should be the highest grade of any now exposed in cuts. Gulch gravels were hydraulicked in Galena Gulch. The bedrock gravel at the head of the hydraulic cut is reported to be good. The lower end of the hydraulic cut was not worked to bedrock. It may be possible to find good panning gravel here beneath recent wash and material caved in.

The terrace gravels were worked in many places. The most extensive placers were the Mekka placers on the north slope of Barney Ford Hill, the Yuba Dam placers on the east side of the Blue one mile north of Breckenridge, the Lomax Gulch and the Banner placers on the west side of the Blue (the first between Sawmill Gulch and Shock Hill, the second between Cucumber and South Barton Gulches) and the Buffalo placers on Ryan and Saltlick Gulches west of the Dillon Reservoir. The gravels vary from 10 feet thick at the Mekka placer to more than 105 feet thick at the Buffalo placers. There were several discontinuous false bedrock levels.

Select panning gravel off bedrock where it is exposed; where it is not, select gravel from the most cobbley sections of the banks. The south end of the Mekka placer cut reportedly still has some good grade gravels.

In the moraine placers in French Gulch at Rich and Weber Gulches select panning gravel from any sorted, cobbley zones (these might be meltwater channels). If these cannot be found, select gravel from places in the banks that have more abundant round cobbles and boulders for these may be places where a greater proportion of stream gravel is mixed with a smaller proportion of glacial debris.

The gold in the valley placers of the Blue, the Swan and French Gulch is concentrated on bedrock and cannot be reached. The gravel in the banks of these streams is probably poor panning at most places except where gulches that have been extensively placered flow into these streams. For instance, the south banks of the Swan at and just below its junction with Galena Gulch are worth prospecting.

LAKE AND CHAFFEE COUNTIES

Lake and Chaffee Counties have many placer showings and three important deposits—California Gulch, Derry Ranch and Cache Creek. The two counties embrace the valleys of the Arkansas River and its tributaries from the headwaters at Tennessee Pass to below Salida. Their western boundary is the Sawatch Range, which is the Continental Divide; their northeastern boundary is the Mosquito Range. The bottom of the Arkansas Valley is broad, with considerable relief. There are notable breaks in slope at the foot of the Sawatch Range on the west and of the Mosquito Range on the northeast.

Glaciers occupied the major western and northeastern valleys. In general, the mountain canyons of the major tributaries were glacially eroded and no significant placers occur in them below the elevations of their lateral moraines. The terminal moraines of these glaciers lie below the mouths of these canyons; however, only two glaciers, those of Lake and Clear Creeks, extended across the Arkansas Valley and diverted the River. Placers are associated with some of the terminal moraines and also occur in some unglaciated small valleys.

Along the Arkansas Valley there are extensive terraces. Four of them are outwash terraces of distinct glacial advances—Early and Late Bull Lake and Early and Late Pinedale but the most extensive and highest, 360 feet above the River, called Terrace No. 1, is a pre-Bull Lake alluvial terrace. It is well exposed south of Leadville, where it extends from Malta to Big Union Gulch. Placers are locally associated with some of the terraces.

California Gulch

Lake County's California Gulch and Georgia Gulch in Summit County were the richest and, considering their size, the most productive placers in the State. The total production of California Gulch is estimated to have been $2.5 million to $3 million. At the reported values of its gold, $17 to $19 per ounce, the production was 132,000 to 176,000 ounces.

In late 1859 prospectors found the first indications of gold in California Gulch at its junction with the Arkansas River, but it was April, 1860, before the rich placers in the upper gulch were discovered. One hundred-foot claims along the gulch amounted to 339. The gravels in the upper gulch were spotty. During the first eight years, the upstream 92 claims are reported to have paid variously from nothing and $10 per man per day to $100,000 in six weeks (at $17 to $19 gold). The claims downstream had paid nothing at that time.

According to tradition, recovery of gold in California Gulch was troublesome because of the heavy "blue sand" that remained with the gold in the pan, rocker, long tom or sluice and that required extra effort to be separated from the gold. Surprisingly, it was not until 1874 that A. B. Wood identified the "blue sand" as cerussite, lead carbonate, and the following year he and Will Stevens discovered the first two of Leadville's great silver-lead orebodies, one on either side of the gulch.

The source of the gold was probably the Printer Boy deposit on Printer Boy Hill. The placer workings do not continue upstream in California Gulch above it, nor are there placer workings in the draws or on the hillslopes on the north side of the gulch. However, there are placer workings in Eureka Gulch (southeast of the site of Oro) and on the hillslope below the Printer Boy mine on the

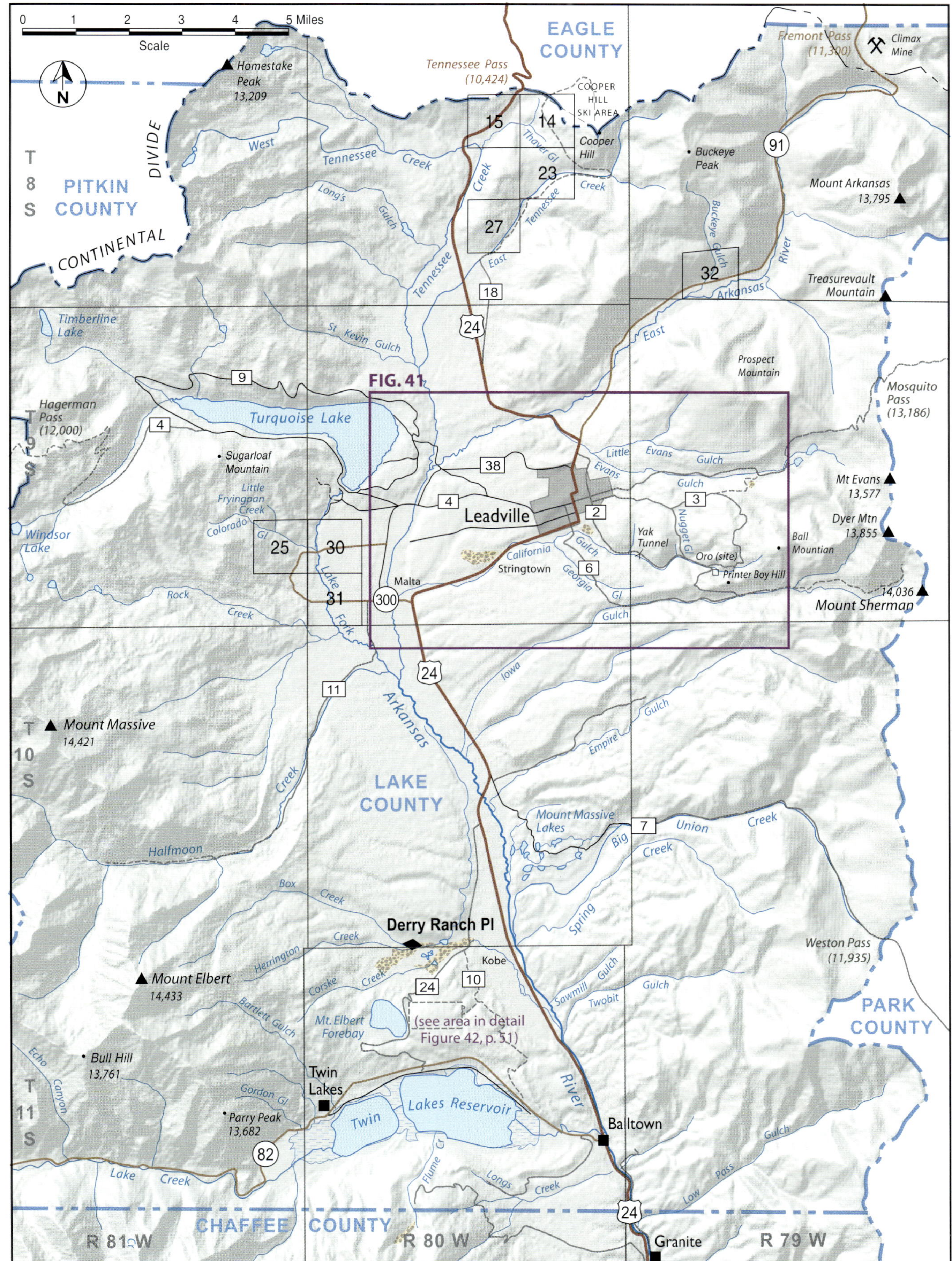

Figure 40. Location map of Lake and Chaffee County placers.

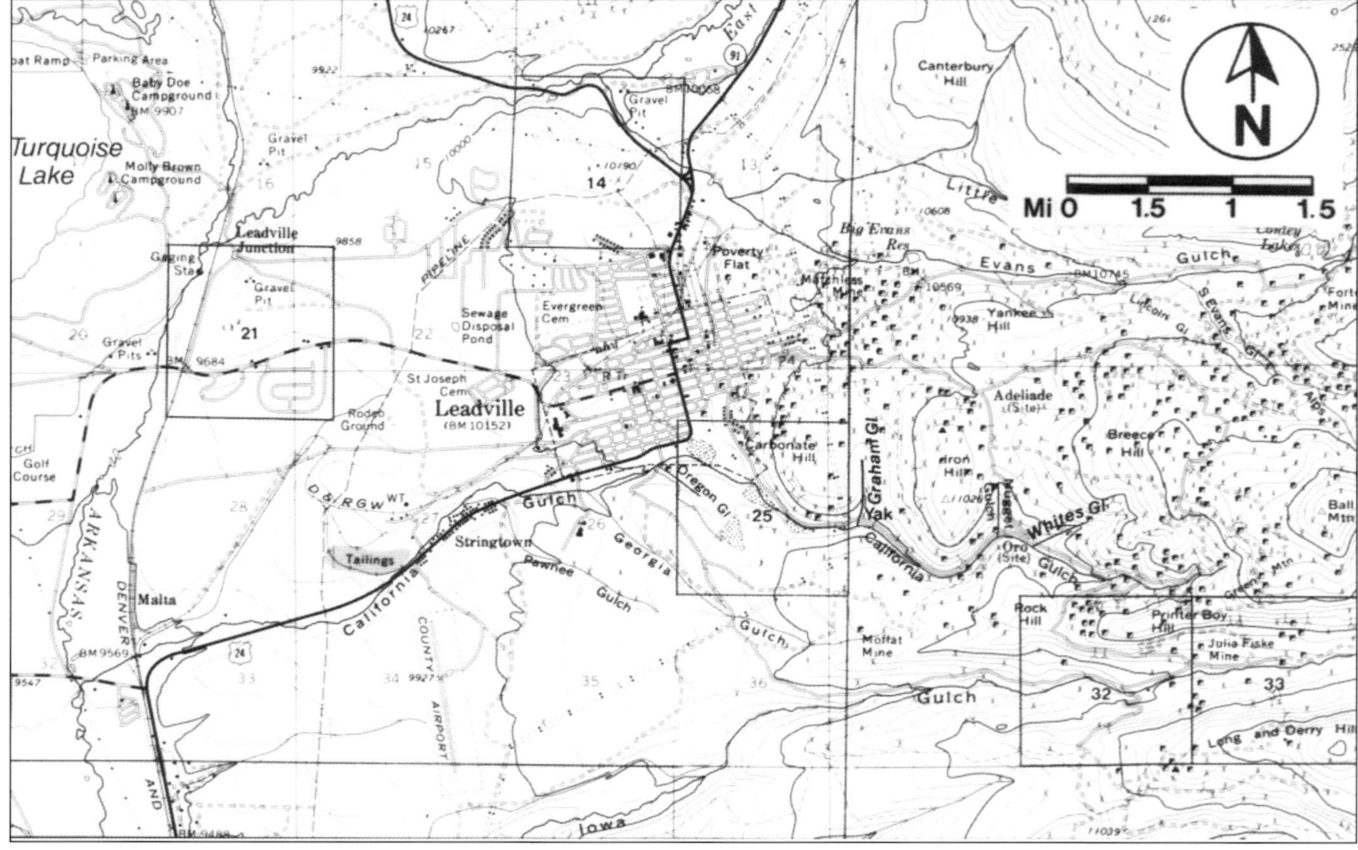

Figure 41. Topographic map of the Leadville area, Lake County. (U.S.G.S. County map Series)

south side of California Gulch. This mine developed a zone of fractured porphyry in which native gold, galena and sphalerite occurred.

California Gulch was placered from the Printer Boy downstream to the south side of Leadville, just above the mouth of Georgia Gulch in the upper gulch the gravels were reported to have been richest at the bend below the site of Oro at the mouth of Nugget Gulch, in the bend about 1,600 feet below Nugget Gulch and in the bend below Graham Gulch near the portal of the Yak Tunnel. From a point about 1,600 feet downstream from the Yak Tunnel the grade of the gravels dropped abruptly. From Georgia Gulch to the Arkansas, although it has been extensively tested, only a little small-scale placering has been done in California Gulch.

From about the Yak Tunnel upstream, the placers were worked to bedrock. Downstream bedrock was too deep to reach and the placers apparently bottomed in compact gravels of Terrace No. 1. In the vicinity of Oro the early miners found a breccia of angular fragments of porphyry cemented with limonite mineral rust. This lay on bedrock and formed a false bedrock above it. The breccia is gold-bearing. Fragments of this breccia can be found in the boulder piles.

Terrace No. 1 gravels occur on the hills to either side of California Gulch, and the Terrace appears to have extended across and buried the gulch at one time. If this is so, it is possible that the downstream bedrock course of old California Gulch is still covered with terrace gravels; that it leaves the present course of California Gulch near the point that the placer workings no longer reached bedrock and near where the grade of the gravels abruptly dropped. If so it may extend, undiscovered, somewhere to the southwest.

The California Gulch deposit is an alluvial placer. If it was once buried beneath Terrace No. 1, the upper part of the gulch, where bedrock was reached in the placers, is older than the Terrace, or pre-early Bull Lake. The lower part, where terrace gravels formed the working bedrock, is much younger and has formed since Terrace No.1 was stripped away.

Upper California Gulch has been extensively placered and there is very little unworked gravel, but the old-timers may have left or spilled some on bedrock from place to place.

If you select panning gravel here, select it from the bottom of banks where gravel is exposed or dig down to bedrock beneath the debris. At Oro drift mining was done and a few—very few—pillars may be left. If you encounter the limonite breccia near bedrock, lift it down to bedrock and pan the material on it, with it and beneath it on bedrock.

A few colors can be washed also from deep gravels from lower California Gulch.

Derry Ranch

The Derry Ranch placer is on Box and Corske Creeks, to the west of the Arkansas River at Kobe station and some two miles north of the Twin Lakes Reservoir. It appears to have been discovered after the 1860s, and was worked by dredging and by a dry land plant for at least 17 years between 1915 and 1950. More than 68,837 ounces of gold have been produced.

The deposit lies in upper Bull Lake outwash and the modern valley floors incised into it along Box and Corske Creeks, on the northern side of the Lake Creek Bull Lake terminal moraine. The immediate source of the gold in this glacio-fluvial placer is Bull Lake glacial moraine, from Bartlett Gulch and from Lake Creek. The original source is probably the Gordon-Tiger group of veins and others on the slopes of Parry Mountain. The gravels worked in the placered area were up to 50 feet deep. Near-surface colors can be found in pits in SW sec. 4 and SE and W sec. 5, T. 11 S., R. 80 W., above the heads of the placered areas.

Cache Creek Park

The Cache Creek placer was discovered in 1859. It is noteworthy for a number of reasons: 1) It was a substantial producer. Most of Chaffee County's production must have come from this placer, so it yielded in excess of $1 million at the old $20.67 per ounce price. No information is available regarding the fineness of the gold. The production had to have been in excess of 49,000 ounces. 2) It was the State's only profitable large hydraulic mine. 3) It was probably the longest operated single placer deposit in the State. From 1859 until 1883 it was worked by individuals and small groups. From 1884 until 1911, when hydraulic mining was stopped by injunction, the mine was operated each season. From 1911 through 1917 placer mining operations decreased, ceasing in 1918. The mine was in operation for nearly 60 years. Subsequently there has been only occasional, small-scale placer mining.

Cache Creek Park is bounded on the north by the Bull Lake terminal moraine of the Lake Creek glacier. The park is bounded on the west by Lost Canyon Mountain and the east by a granite ridge. This ridge rises only slightly above the level of Cache Creek Park to the west, but the Arkansas River has cut a deep canyon below its crest on the east side. Within the park, Long's Gulch is on the south edge of the Lake Creek moraine. Oregon Gulch is its southern tributary. Still farther south Gold Run and Cache Creek once flowed across the Park; the placer cut has destroyed their valleys. Cache Creek changes its name at the foot of Lost Canyon Mountain; on the mountain it is called Lost Canyon Creek.

The faces in the placer cut are up to 61 feet high. At many places they have a distinct bench about midway in the exposed gravel. This was a false bedrock. The cobbles and boulders above it are sounder than those beneath it, suggesting that the lower ones have been weathered for a much longer time. The gravel below the bench is lower Bull Lake outwash; the gravel above it is upper Bull Lake outwash. That the bench is not present in all the faces shows that the upper Bull Lake outwash streams did not flow over all the Park but were restricted to limited channels.

The placer was worked in two "lifts;" the working bedrock of the first lift was about the level of the benches. The bedrock of the second lift, the present floor of the cut, is soft, deeply weathered granite on the east side of the cut and gray and orange clay in the center and west side. In Bull Lake time the glaciers of Lake Creek and Clear Creek pushed across the Arkansas River valley and up its lower east slope. The river was diverted off its valley floor and it cut its present canyon in the granite. From the west edge of this canyon the granite surface slopes downward beneath the moraines toward the mountains on the west. This can be seen in the placer cut, where the granite bedrock slopes downward as the cut is entered from the east. From midway in the cut to its western sides, gray and orange clays, sands and conglomerates, called the "Lake Beds," overly the granite. The pre-Bull Lake valley floor of the Arkansas River must have been cut in these beds and the lower Bull Lake outwash gravels were deposited on them (they form the bed-rock of the Derry Ranch placer to the north of Twin Lakes).

Cache Creek Park had two sources of gold: The principal source lay to the west or southwest—probably Lost Canyon Creek—and the moraines of Lake Creek also provided some gold.

Cache Creek Park was an outwash terrace of lower Bull Lake age, 155 feet above the Arkansas; its gravels contained gold from both sources. In upper Bull Lake time these gravels were reworked by streams which brought in more gold from both sources and additional gravel. Gold Run (destroyed by extensive placer workings) and Cache Creek must have been sites of such reworking and deposition of upper Bull Lake outwash; their valleys were destroyed by the placer mining. Long's Gulch and Ritchey's Patch were the sites of similar reworking and Long's Gulch is still floored with upper Bull Lake outwash. Long's Gulch probably did not receive gold from Lost Canyon as did the other areas and so the placer deposits in it are much less important. Gold Run and Cache Creek continued to receive gold from Lost Canyon after Bull Lake time as Pinedale and more modern streams reworked some of the upper Bull Lake outwash.

GUNNISON COUNTY

North Taylor Park and Bertha Gulch have been only small placer gold producers but they are of special interest because their placers carry some heavy minerals besides gold and magnetite. They are in northeastern Gunnison County in the Upper Taylor River area. Several other placers there are mentioned in the Challenging Areas for Panning section (see pages 66 and 67).

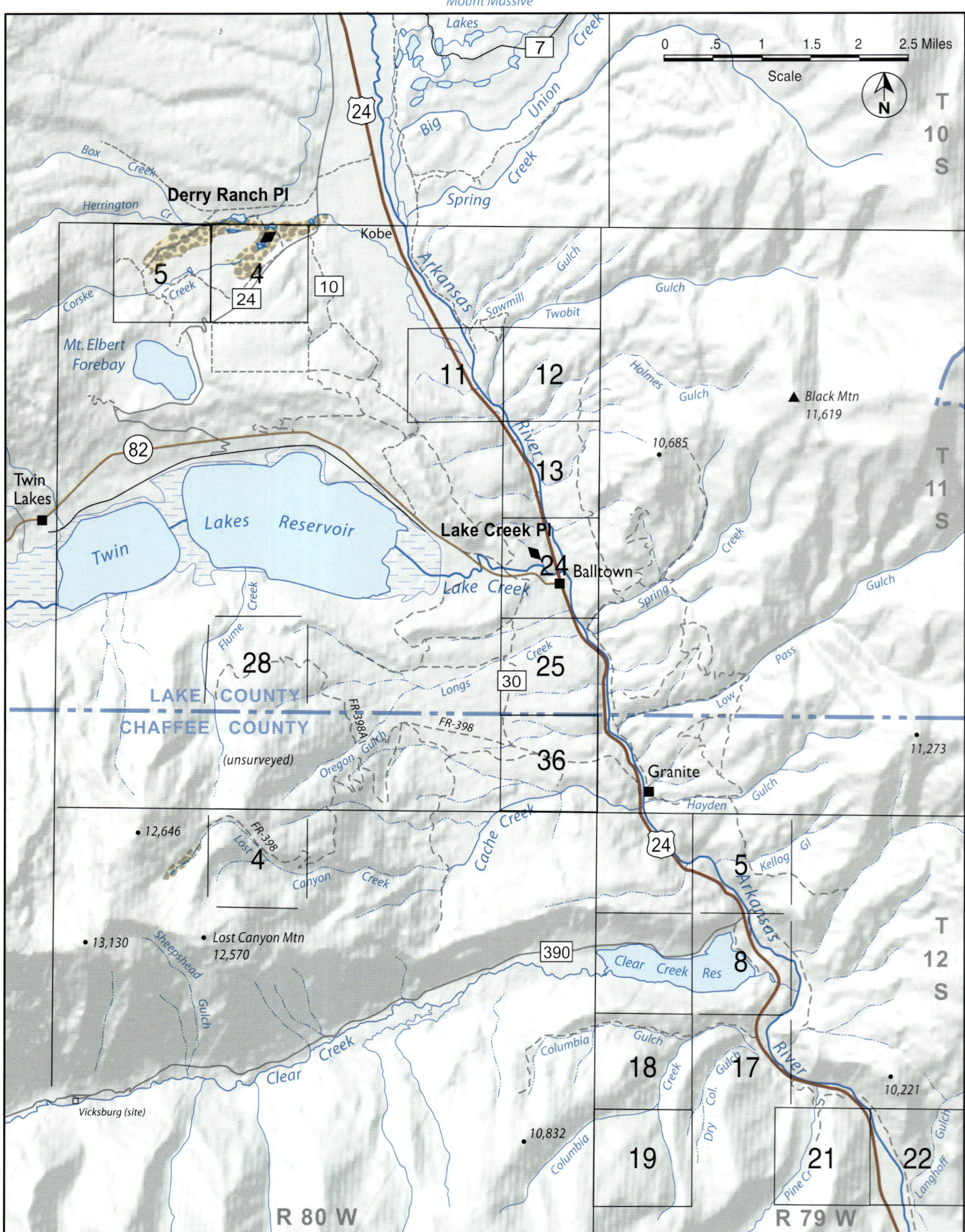

Figure 42. Location map of Derry Ranch and Cache Creek placers, Lake and Chaffee Counties.

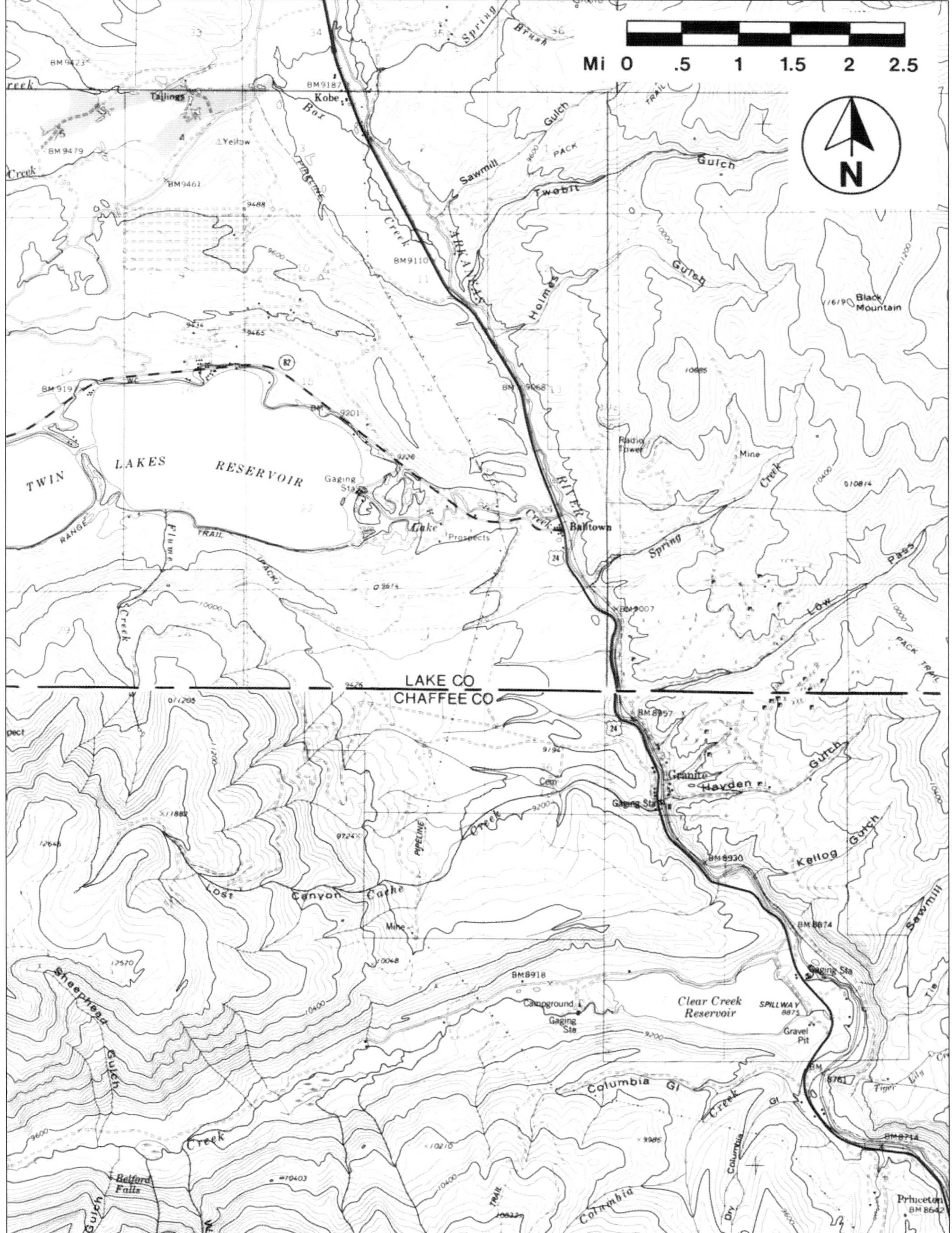

Figure 43. Topographic map of Derry Ranch and Cache Creek area, Lake and Chaffee Counties. (Modified from U.S.G.S. topographic maps)

Figure 44. Location map of Gunnison County placers.

The first reports of mining in North Taylor Park, at Kent Gulch, were in 1867; probably work began in the nearby gulches about the same time. The first report of placering in Bertha Gulch was in 1881. Desultory mining continued at both places through the 1880s. In 1896 the Taylor Park Placer Company completed its ditch, bringing water from the Taylor River to Little Taylor Gulch, and placering continued in North Taylor Park for a few years afterward. There was a little activity again in Depression times.

Most of the work in these two areas was manual—rockers, booming and shoveling-in augmented by small-scale hydraulicking where water was available. The Placer Company began construction of a bedrock flume at the Taylor River to work Illinois Gulch. The work was abandoned in 1899 when the bedrock gradient was found to be too low to operate a flume. (This was also the final result of other bedrock flume projects in the district at Little Taylor Gulch and Union Canyon.) The only other large scale mining attempted was the operation of a hydraulic elevator in Illinois Gulch in 1898.

North Taylor Park

These placers lie in Illinois and Pieplant Creek and their tributaries' valleys below 9850-feet elevation on Pieplant Creek and below 9750 feet on Little Taylor Gulch. Taylor River and Illinois Creek have broad, outwash filled valleys. From place to place along Illinois Creek there are terraces 85 to 95 feet above stream level.

Placering was attempted in the outwash plain at the mouths of Illinois Creek and of Little Taylor Creek, but the workings appear not to have reached bedrock. Placers have been worked from place to place in most of the tributary gulches between Illinois and Pieplant Creeks, in a few spots on the terraces and on the hillslopes and in ravines on them. The gulch placers and those on the slopes and ravines appear to have been profitable; the terraces were little worked and apparently offered encouragement to the miners in only a few places. In the gulch placers the gold-bearing gravels were covered with barren debris which thickened upstream. The gulches were worked to where it became uneconomical to move this cover.

On the hill slopes and in the ravines the gravels were usually covered, but with no more than three feet of saprolite-granular, disintegrated granitic rock debris.

The gold found in these placers is reported to be small flakes, wires and, rarely, small nuggets. With the black sand in pan concentrates there frequently is minor columbite-tantalite, monazite, zircon and garnet. Columbite-tantalite is an ore of tantalum; it appears as black sand, resembling magnetite but is non-magnetic. Monazite is a thorium-bearing mineral; it appears as blackish to greenish grains and is slightly radioactive. Zircon appears as yellow to orange grains and garnet most commonly as red to orange grains. These minerals are not abundant in the concentrates; they have never been produced commercially.

The source of these other heavy minerals is the granites and the pegmatites in the area, but the source of the gold has never been identified. The placers are widespread; this suggests that there are a number of sources. The placers are below the Taylor Park placer ditch. Whether there were no gold-bearing gravels above it or that none were worked above it because of lack of water is unknown.

Gravels for panning can be found in some old workings in the gulches and ravines. Debris on bedrock in old hillslope placer areas should yield some colors and any gravel found on the slopes should be panned. The terraces are probably low grade. Secs. 27 or 28 should be good places to start.

Bertha Gulch

There is an extensive hydraulic cut in the fan gravels at the mouth of Bertha Gulch and small cuts from place to place upstream across sec. 1, T. 15 S., R. 82 W.

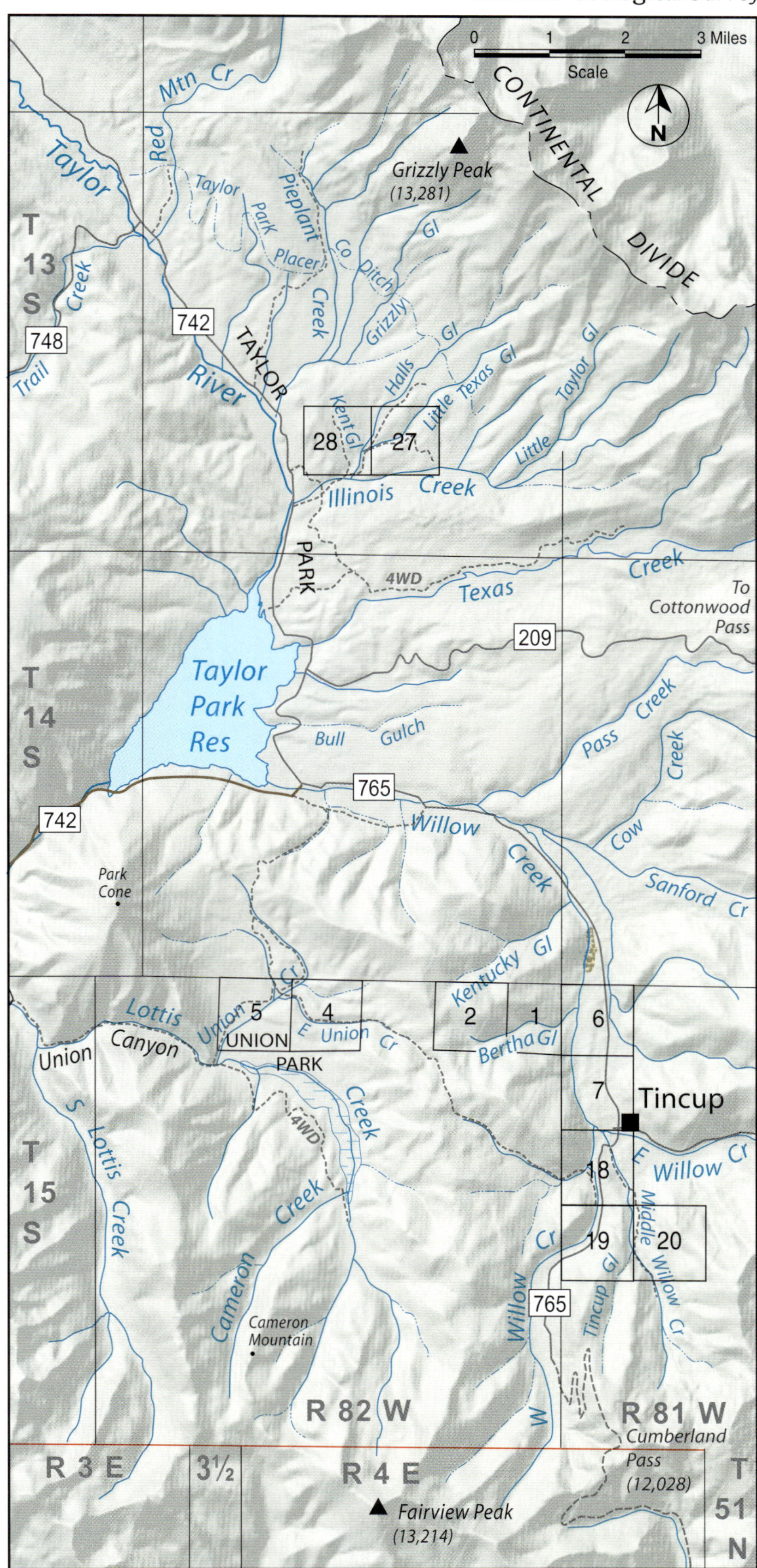

Figure 45. Location map of North Taylor Park and Bertha Gulch area placers, Gunnison County.

The gravels of the hydraulic cut reportedly were better than those upstream; the best of the upstream gravels were just west of the center of sec. 1.

Scheelite, an ore of tungsten, occurs in the heavy minerals of the upstream gravels. Its recognition in pan concentrates led to the discovery of gold-scheelite-quartz veins on the north slopes of the gulch in sec. 2. These veins did not prove to be economic, but they are a bedrock source for the gravel here. Some placer gravels are still exposed in the hydraulic cut and a few places upstream.

Similar but much smaller placers were worked in Kentucky Gulch, which lies 1 to 1½ miles to the north. These placers are ones that you may have to prospect some to come up with a good pan of gravel.

CHALLANGING AREAS FOR PANNING

CHERRY CREEK DIVIDE AREA

Arapahoe, Denver, Douglas, and Elbert Counties

This area includes the sites of the gold discoveries of 1858, which led to the "Pikes Peak or Bust" gold rush of the following year. The first worthwhile deposit found was on the South Platte River, in Denver County, probably in sec. 21 and 22 T. 4 S., R. 68 W., (near Overland Park) between the present 8th and Jewell Avenues. The most productive deposits found later were on Big Dry, Newlin and Russellville Gulches in the Cherry Creek drainage in Douglas County and on Ronk (now called Gold Creek) and Gold Run Gulches to the east of Russelville Gulch in Elbert County.

The placers were pretty much abandoned by the end of 1859; the miners left for the richer placer areas discovered elsewhere that year. However, there has been both gulch and drift mining from time to time since, especially individual and small- scale operations during Depression times.

The gulches in which the placers lie drain the Cherry Creek Divide, which is capped by the Oligocene Castle Rock conglomerate. (This formation forms "Castle Rock,' overlooking the town of the same name.) To the northeast of Castle Rock town the conglomerate has been eroded by the streams and now is represented only by remnants capping buttes along the divide. Many hillslopes have residual pebbles and cobbles on them, derived from the erosion of the conglomerate. There are terraces or benches at many places along the streams. Along the South Platte the terraces are 40 feet above the level of the river. These benches extend into Cherry Creek and its tributaries, their height diminishing upstream.

Gold has been found in the terraces and in the stream beds. Some testing has shown that, at least locally, gold in the bench placers lies on or near bedrock but along the stream beds it is nearer surface. In most of the productive gulches gold values are spotty. Along Happy Canyon and Big Dry Gulch reportedly gold is widespread but values are very low. In some gulches, among them Badger, Oak and Cottonwood Gulches and the Plum Creek drainage, no gold has been found.

The Castle Rock conglomerate contains a little gold at many places. In upper Newlin Gulch near the forks some small, lens shaped beds of gravel were found in it that reportedly contained up to one ounce of gold per ton in individual samples. ***This area is now partially inundated by Reuter-Hess reservoir.*** After their discovery in 1895, these were worked in several small drift mines. The lenses were 5 to 40 feet above modern stream levels, probably near the base of the Castle Rock formation.

The gold in all the placers was typically fine-grained-10 to 50 colors per milligram. Miners described it as "almost flour gold" and complained of the difficulty of its recovery from the clayey gravel matrix. Even so, colors up to one-sixteenth-inch diameter were occasionally found. The gold is unusually pure, characteristically about 990 fine.

The immediate source of the stream and terrace placer gold is the Castle Rock conglomerate and fossil stream placers in it. That the source is fossil stream channels does much to explain the erratic distribution of gold among the gulches and in each of them.

Access to the placer gravels is difficult because of the many developments in the area. There is very little water and the gravels must be packed out. ***Remember to get permission to enter and to leave gates as you found them, open or closed!***

The gravels of Cherry Creek and of the South Platte below the mouth of Big Dry Gulch contain some gold but are very low grade and spotty. Some samples of the Castle Rock conglomerate, particularly of coarse, cobbley sections that may be old stream channels, will show fine colors. Perhaps the best stream areas are Newlin Gulch at and below its forks and Russellville, Ronk (Gold Creek) and Gold Run Gulches and other gulches in T. 8 S., R. 65 W., near and below the base of the Castle Rock conglomerate.

SOUTH BOULDER CREEK AND TRIBUTARIES

Boulder and Gilpin Counties

From the junction of Beaver Creek with South Boulder Creek for about 3.5 miles downstream South Boulder Creek is the boundary between Boulder and Gilpin Counties. The Deadwood Diggings, the third of the economic gold discoveries made early in 1859 that maintained the Colorado gold rush, was on South Boulder Creek near the mouth of Beaver Creek. Gold-bearing gravels probably extend downstream for some distance, but little evidence remains of early day activity there.

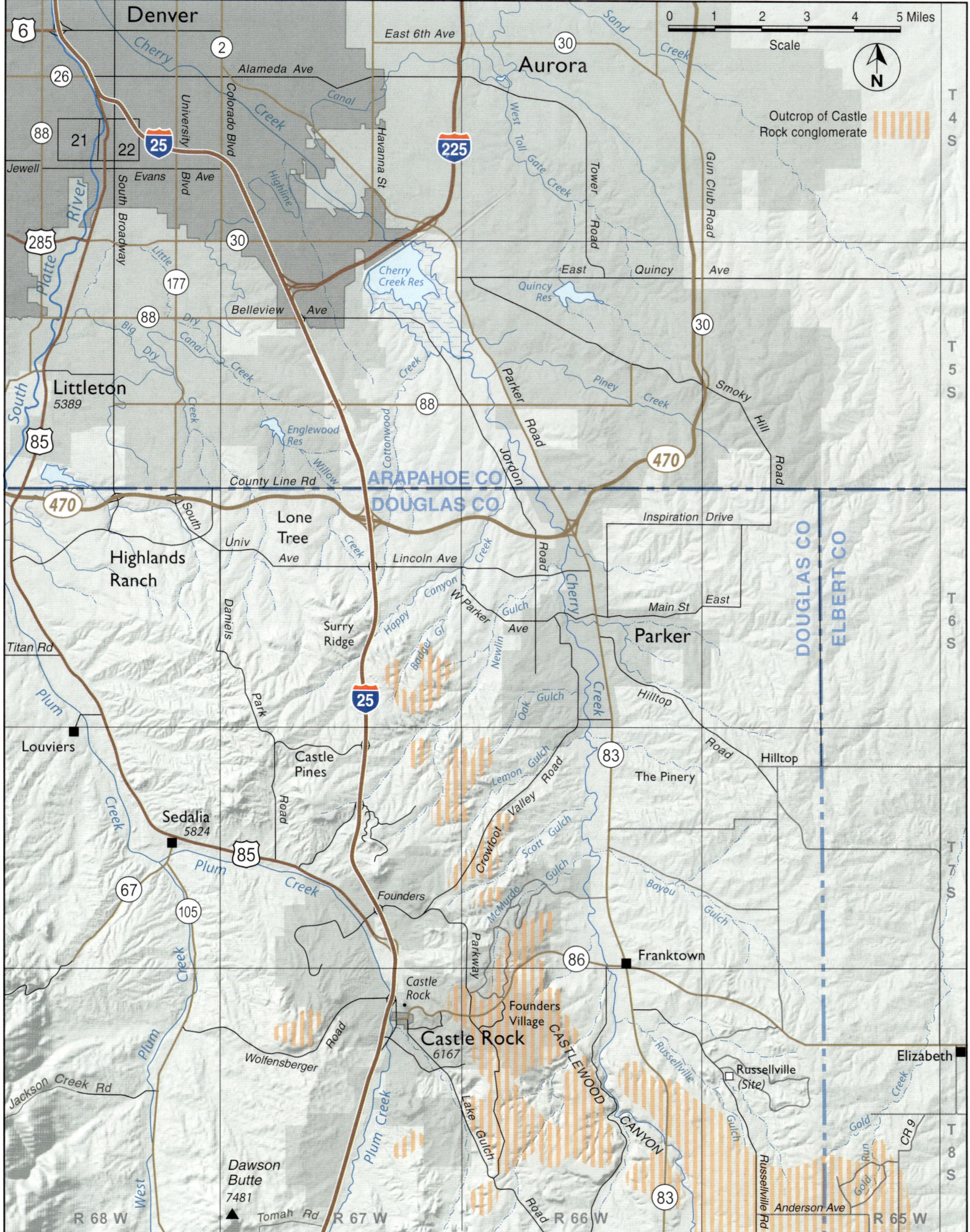

Figure 46. Location map of the Cherry Creek Divide area placers, Arapahoe, Denver, Douglas, and Elbert Counties.

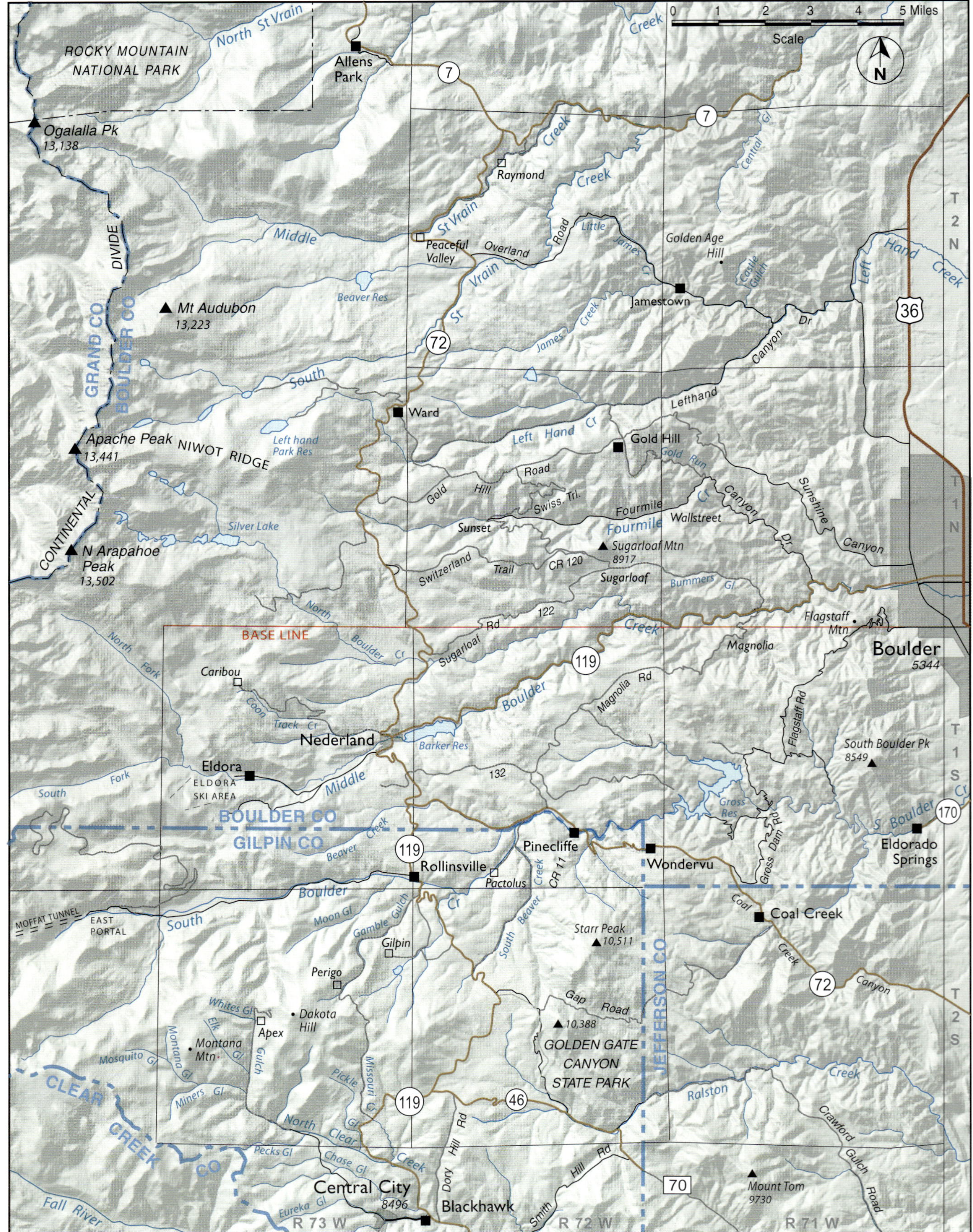

Figure 47. Location map of placers in South Boulder Creek and tributaries area, Boulder and Gilpin Counties; and the Upper North Clear Creek placers, Gilpin County.

Gilpin County

Within Gilpin County the gravels of South Boulder Creek are gold-bearing from the mouth of Moon Gulch to Beaver Creek. On South Boulder Creek there are two parks where the valley floor widens, one centered on Rollinsville and the other centered on Pactolus Station and extending to the Deadwood Diggings area. Both were worked with dry land washing plants in the 1930s. Earlier placers were worked in Lump Gulch from near where Colorado 119 approaches it from the south upstream for about three-quarters of a mile and in Gamble Gulch from near where Colorado 119 crosses it upstream to the site of the old village of Perigo. These deposits are stream placers. Some gravel still can be found in their banks from place to place.

UPPER NORTH CLEAR CREEK

Gilpin County

Small placers were worked along North Clear Creek and Missouri Gulch. Those in North Clear Creek are just above the mouth of Miners Gulch and from a quarter of a mile below the mouth of Mosquito Creek upstream along Montana Creek to near its head. Along Missouri Creek there were placers near North Clear Creek, but the most extensive workings were upstream from where Colorado 119 crosses Missouri Creek near Missouri Lake. The workings extend from the highway upstream for about 0.7 miles There are some gravels remaining from place to place on North Clear Creek and reportedly above the old placers on Missouri Creek. Placers were probably worked on Pecks, Miners and Whites Gulches and on Elk Creek. These gulches might be worth exploring.

LOWER CLEAR CREEK AND SOUTH PLATTE RIVER

The gravels in the terraces overlooking the creek eastward from the gap between the Table Mountains are gold-bearing in many places and locally to some distance from the creek. Such gravels are reported in secs. 2, 4, 12, 10, 19 and 24, T. 3 S., R. 69 W. Although some of these sites are on Ralston Creek, the gold is from terrace material deposited by Clear Creek. In general the grades are low, but occasionally gravels immediately overlying bedrock have many fine colors. Gold has been produced from terrace and valley floor gravels in Adams and northeastern Jefferson Counties as a product of sand and gravel washing. The grades in Adams County were very low, but gold recovery was profitable since the gold was a very low cost by-product of already commercial gravel washing. Any pit banks in the elevated gravels west of Wadsworth avenue and banks in lower Ralston Creek are favorable sources of gravel for panning.

Less favorable but with some chance are point bars in Clear Creek east of Wadsworth Avenue to the South Platte River and on the Platte downstream. Gold has been found in Platte surface gravels for many miles below the mouth of Clear Creek. However, the farther downstream from Golden, the finer and the scarcer will be the gold and the more difficult to recover. Again, this is private property, so permission should be obtained.

BOULDER COUNTY

Placer gold has been produced at the "Deadwood Diggings" on South Boulder Creek (see page 56), at the "Jefferson Diggings" on North Beaver Creek about four miles above South Boulder Creek, along Gold Run south of Gold Hill, along Fourmile Creek from place to place from Boulder Creek to near Sunset, "near Sugarloaf" (probably Bummer's Gulch), on Coon Trail Creek between Nederland and Caribou, on Castle Gulch and just below its mouth on James Creek, and at the head of Central Gulch northeast of Jamestown. (The last two gulches drain the slopes of Golden Age Hill, below the vein of the same name.) Placer gold is reported in the gravels of Boulder Creek from the west side of Boulder to the mouth of Fourmile Creek and on the North St. Vrain north of Allens Park. The gravels at both these last two sites are probably extremely low grade.

CHAFFEE COUNTY

Lost Canyon

The Lost Canyon placers lie on the northeast slope of Summit 12,570 in secs. 4 and 5 (unsurveyed), T. 12 S., R. 80 W., two miles west of Cache Creek Park. The placers lie in a basin between 11,250-feet and 12,100-feet elevation. The placers extend from the forks of Lost Canyon Creek upstream along the northern, main branch more than a mile and downstream a short distance.

In 1860 prospectors returning from Taylor Park discovered this deposit. Reportedly they were lost when they found it, hence the name. Summit 12,570 lies at the eastern end of a high ridge 10 miles from the Continental Divide, which forms the eastern limit of Taylor Park. That travelers coming east from Taylor Park would pass by this mountain suggests that they were truly lost or, more likely, that in the then trackless country it was easier to travel above timberline than in the forested valleys.

Operators have told me that although gold is concentrated in a pay gravel on bedrock, this gravel varies from a few inches to 15 feet thick. The pay gravel contains angular rock fragments, with fewer large fragments than the mantle over it, and the largest fragments in it are less than 18 in diameter. It contains little clay and is easy to wash. The gold is coarse and rough; gold is frequently attached to quartz fragments.

This deposit is unusual in several ways: 1) Despite the high elevation of its head, Lost Canyon was not glaciated in Illinoian and Wisconsin time, so the placer is old and has developed since before Illinoian time.
2) Below the placers the valley changes from an open

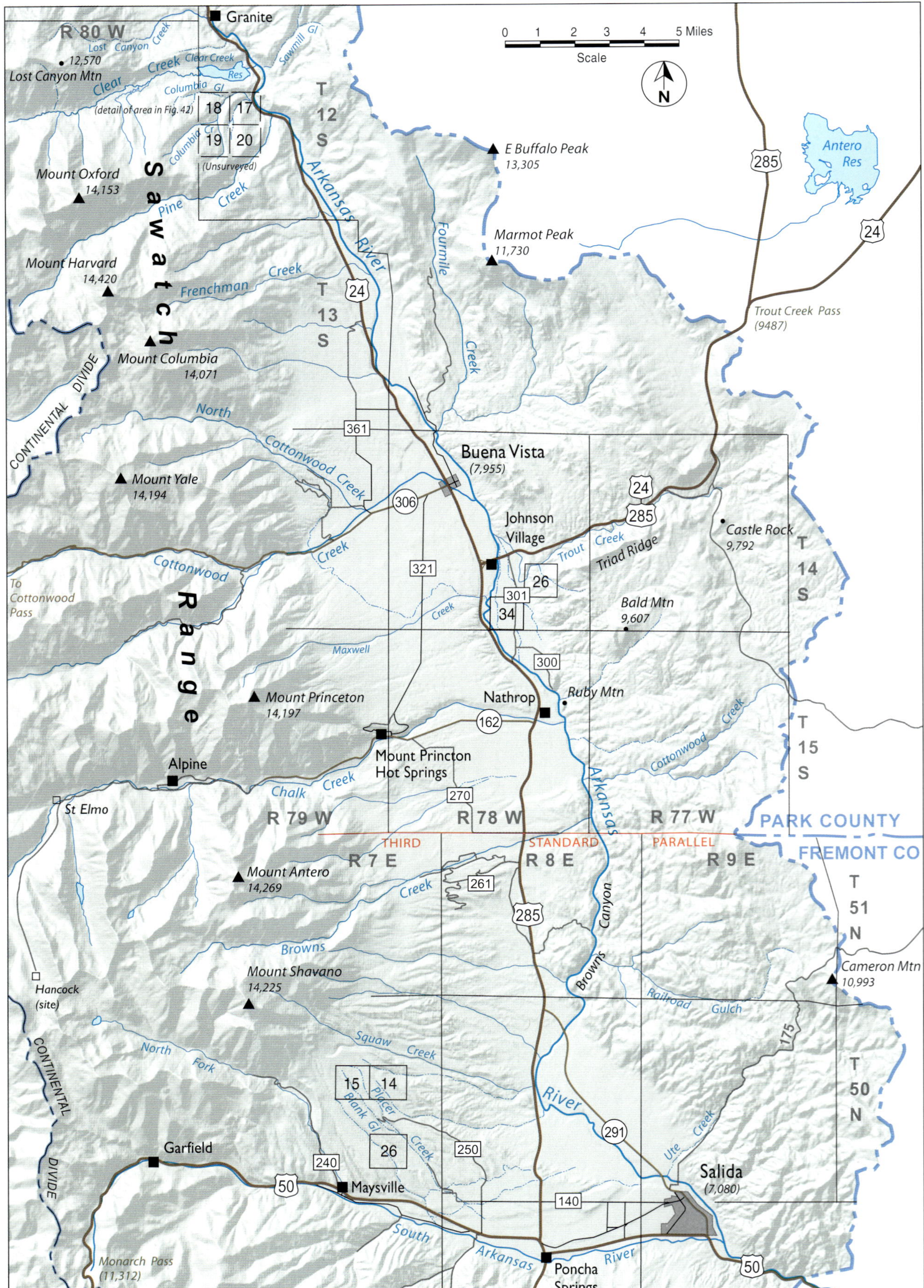

Figure 48. Location map of placers in Lost Canyon, Columbia Gulch, Mt. Shavano, and Trout Creek, southern Chaffee County.

basin to a very steep, narrow valley incised 100 to 150 feet into the mountain slope, and just above the level of Cachè Creek Park it changes again to an open, gently sloping valley. These lower sections of the valley appear to have been well prospected but never worked. Reportedly, the pay disappeared a short distance below the forks in sec. 4. It seems reasonable that upper Lost Canyon was a source of some of the gold in southwestern Cache Creek Park, but no placer gold has been found between them. 3) The source of the gold in Lost Canyon has never been recognized. It must lie in the upper basin, but whether one or several good bedrock deposits or a widespread, low-grade occurrence is unknown.

Select material for panning like the pay gravel described above. If there is none, take the material as deep as possible in the cuts. There is some suggestion that mantle away from the stream may contain gold from place to place. Should you wish to try this "long shot," prospect the upper and northern parts of the basin, where mantle is thinner. By mid-summer water is scarce here and you must "either pack water up or gravel down."

Columbia Gulch

Placers have been worked on Columbia Gulch and Columbia Creek in secs. 17, 18, 19 and 20, T. 12 S., R. 79 W.

Mount Shavano

There are placer workings on Placer Creek in SWSW sec. 14 and ESE sec. 15, T. 50 N., R. 7 E. and at the junction of Placer Creek with Blank Gulch in sec. 2.

The placers are in valleys incised in a piedmont alluvial plain (a plain formed at the foot of mountains by the coalescing of gently sloping alluvial fans and stream deposition between them) at the foot of Mount Shavano. The thickness of the gravel beneath the plain is unknown, but across the South Arkansas valley the gravel beneath a similar plain is more than 200 feet thick. The "bedrock" at the placer deposit is probably undisturbed piedmont gravel. This plain is pre-Illinoian in age. Blank Gulch and Placer Creek began to be incised in it sometime after its deposition and the placer deposits have developed since that time.

The gold produced at these placers is said to have been heavy and rough and the placer gravel to contain a great deal of black sand. The original source of the gold apparently is a mineralized area high on the southeast face of Mount Shavano, where reportedly some eluvial placers were worked years ago.

Select gravels for panning at either end of the workings on Placer Creek. If you like the area, the placer channels (if any) between the Placer Creek workings and the bedrock deposits on the peak have never been found.

Placers are reported on upper Browns Creek, five or six miles to the north in T. 51 N., R. 7 E. If this is true, some deposits yet to be discovered may exist at the foot of the mountain and the top of the alluvial plain in the gulches between Blank Gulch and Browns Creek.

Trout Creek

Placers were worked on Trout Creek " one mile from the Arkansas River," probably in secs. 26 and 34, T. 14 S., R. 78 W.

CLEAR CREEK COUNTY

Alice/Yankee District

The Alice mine, marked by a large cut, is 0.6 mile southwest of Alice in sec. 3 (unsurveyed), T. 3 S., R. 74 W., at approximately 10,300-feet elevation. Yankee Hill lies 1.8 miles northeast of the Alice Mine in sec. 36, T. 2 S., R. 74 W., and reaches 11,239-feet elevation. The sites of the ghost towns of Yankee and Ninety-four were near the summit of Yankee Hill. From the Alice mine to Yankee Hill there are several bedrock ore deposits that were discovered and first worked in the early days. The deposits were mostly narrow veins, but at the Alice mine and a few less important ones they were stockworks (large fractured and broken zones in which some or all the fractures are mineralized, forming a "spider web" of generally knife edge veins). These deposits contained quartz with some gold-bearing pyrite; there were few other ore minerals. Cirques and U-shaped valleys show that Silver Creek and Fall River valleys were occupied by alpine glaciers. But the slopes around the Alice mine were protected and little glaciated, and the south shoulder of Yankee Hill overlooking Silver Creek and Cumberland Gulch projected above the ice. So the deposits' outcrops, perhaps weathered since pre-Illinoian times, were not eroded away by the Bull Lake and Pinedale glaciers and the development of the placers continued until modern times. The unweathered, unoxidized material from the veins was at most places too low grade to mine profitably, even in the early days. The stockwork deposit of the Alice mine could be worked on a larger scale and the unoxidized material from it was mined profitably in the 1930s. Any placers that may have formed along Silver Creek, itself, or Fall River were destroyed by the glaciers.

The oxidized material beneath the outcrops had been enriched and loosened by weathering processes. This material was readily and profitably placered by hand methods within a few years—Yankee and Ninety-four were almost ghost towns by 1870. The Alice deposit was ground sluiced in 1882 and hydraulicked in 1883. The outcrop material there was three to six feet of rocks and soil above two feet of yellow earth that was in turn above about three feet of white and gray, clayey shattered bedrock containing quartz crystals and pyrite. The yellow earth reportedly "gave a nice string of colors in the pan." The vein outcrops that were placered on Yankee Hill were eluvial placers. The outcrop of the Alice deposit was also an eluvial placer, but the placers that continued for a short distance downslope were colluvial placers.

Although these placers have long since been worked out, careful prospecting might still yield some "dirt" for panning. ***Be careful around the old mine workings and the edges of the Alice pit and don't enter the workings!***

North Empire District

The "Silver Mountain ore zone" lies in the west sec. 21, T. 3 S., R, 74 W. (see Fig. 33), on the west side of North Empire Creek and about one mile north of U. S. Highway 40. Here there is a large zone of shattered rock, strongly mineralized with pyrite and gold-bearing chalcopyrite. The outcrop material was thoroughly weathered, the pyrite and chalcopyrite leached away and the gold liberated, forming an eluvial placer over the ore zone. There were colluvial placers below it and stream placers on North Empire and Empire Creeks. These continued to Empire but persisted no further downstream, North Empire Creek drainage was not glaciated. Empire Creek (below the ore zone) was covered by the Bull Lake glacier that came down Clear Creek valley but not by the Pinedale glacier. So these placers developed over long periods, but any placer deposits there may have been at Empire and downstream were swept away.

The North Empire placers were discovered in 1862 and 1863 and worked for three or four years thereafter.

COSTILLA COUNTY

The placers of the Grayback Mining District, north of Russell in T. 28 S., R. 71 W. (unsurveyed), are probably the first worked in the State. There is a dredge cut 2,200 feet long on Placer Creek in sec. 36, its downstream end about a mile north of Russell. Placer Creek and its terrace gravels were worked from place to place from Giant Gulch to Strawberry Creek in secs. 25 and 26. Officers Bar, the site of the first mining activity, is a large terrace remnant between Grayback and Strawberry Gulches, on the east side of Placer Creek in NE sec. 26. There are a few small placer workings in Gray-back Gulch from its mouth on Placer Creek in sec. 26 to its junction with Buckskin Gulch in sec. 13, and in Spanish and Giant Gulches. There is gold-bearing gravel in Buckskin Gulch. Placers have also been found and worked from place to place along Willow Creek for two miles up from its junction with Sangre de Cristo Creek in secs. 18, 19 and 30, T. 28 S., R. 70 W.

In the 1850s soldiers from Fort Massachusetts worked Officers Bar in a small way. The Fort, which was on Ute Creek about six miles north of the present site of Fort Garland, was built in 1852 and abandoned for Fort Garland in 1858. Chinese miners are said to have reworked Officers Bar but there are no records of subsequent placering until 1876 and thereafter not until 1896. Beginning the latter year there was considerable small-scale activity for a few years. A small dredge operated in 1910 and in 1911, until it sank. During the 1930s a few individuals placered here.

In the southwest part of the district there are indurated high gravel deposits, called the Grayback Wash, in irregular large remnants from 250 to 300 feet above the level of Placer Creek. The bodies of Grayback Wash on the west side of Placer Creek contain no gold, but one on the east side, which extends from Spanish Gulch to the head of Giant Gulch is gold-bearing, reportedly carrying 3 to 10 coarse colors per pan. The gold ranges from coarse to fine and is commonly scaly. The gravels of Spanish Basin, on Spanish Creek, are derived from the Grayback Wash.

Lower gravel terraces are conspicuous in Placer Creek valley and both branches of Middle Creek. From Officers Bar downstream to Mill Creek there are terraces at two levels, 50 and 90 feet above the creek. The terraces upstream from Officers Bar on Placer Creek are not gold bearing. From Middle Creek downstream the terraces on the west side of Placer Creek valley did not pay to work. The terraces on the east side of the valley are gold bearing; Officers Bar was the richest and most extensive of them. These terrace gravels and those of the stream floor were worked from place to place from Officers Bar to the mouth of Giant Gulch. The gold produced ranged from flour gold to round flakes one-eighth-inch across, with occasional nuggets to one-tenth ounce.

Most of the gold was found in gravels within a few feet of bedrock. There is abundant magnetite in the gravels. The gravels of Buckskin Gulch and Willow Creek contain similar gold.

The distribution of the gold-bearing gravels suggests two sources for the gold—low-grade bedrock gold deposits on Grayback Mountain and the Grayback Wash. Of the various remnants of the wash, only the one on the southwest slope of Grayback Mountain contains gold. Consequently, the gold deposits on Grayback Mountain are probably an original source of the gold in the wash as well as the gold in the placers.

Select gravel for panning from any of the places mentioned above. Usually in this district the more sedimentary and volcanic rock fragments a gravel contains, the more likely it is to contain gold. Gravels with predominantly metamorphic fragments are usually barren. Magnetite is not a guide; both barren and gold-bearing gravels contain abundant magnetite. The gravels in many places are covered with a few feet of soil or fine alluvium which is barren; be sure that your sample has some coarse fragments. Along Placer Creek select your samples from the east side. Although the Grayback Wash is indurated, at the foot of its outcrops material weathered from it may accumulate.

DELTA COUNTY

Some small placers have been worked on terraces beside the Gunnison River on the north side of the river near Hotchkiss in southwestern T. 14 S., R. 92 W., and in SW sec. 15, T. 15 S., R. 97 W.

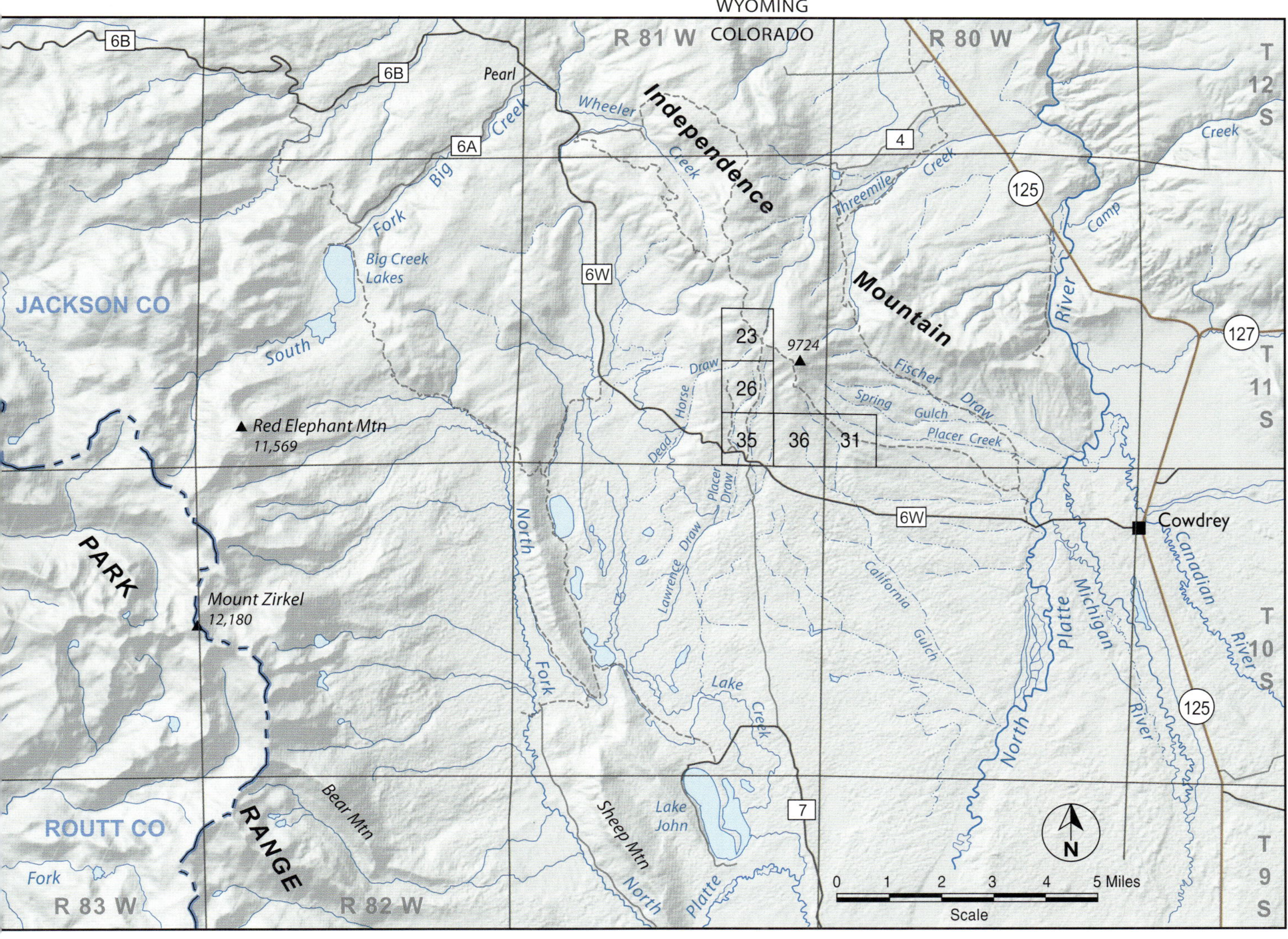

Figure 52 con't. Location map of Independence Mountain placers, Jackson County.

ROUTT COUNTY

Hahns Peak

In the Hahns Peak District there are two principal placer areas: Poverty Bar on Deep Creek northwest of Hahns Peak village in sec. 20, T. 10 N., R. 85 W, and Ways Gulch to the east of the village. There are smaller placers in Little Red Park on a branch of King Solomon Creek north of Twin Mountain and in the vicinity of Columbine. A little mining and some testing has been done along Willow Creek.

The Hahns Peak placer area was discovered by a Captain Way in 1864. In Empire he told John Hahn of his discovery. In 1865 Hahn and W. A. Doyle went to the area and prospected it. They returned with 40 men in 1866, built cabins at the site of the present village and placered through the summer. They established the mining district and named it and the peak for Hahn. Facing an early winter, the 40 men returned to Empire, leaving Hahn and Doyle at the camp. In the spring of 1867 the two men needed supplies. They headed for Empire, and in a blizzard during their journey Hahn died.

Although the district has been only a modest producer, it supported some activities each year from 1866 through at least 1909. By 1876, 20 miles of ditches had been completed, bringing water from the Elk River to the placers, and hydraulicking was to begin. In 1907 a very small dredge worked a few months on Willow Creek near its junction with Ways Gulch. In 1931-39 there were from 1 to 20 small operations each year. There have been sporadic small operations since.

Hahns Peak is an igneous stock intruded into the sedimentary rocks which surround it. These sedimentary rocks include shales, sandstones and conglomerates. No important bedrock gold deposits have been found in the Hahns Peak district. The porphyry which makes up the stock and the sedimentary rocks immediately adjoining it carry varying, but small, amounts of gold and free gold has been found in porphyry from near the contact. The placer gravels contain predominantly sub-angular porphyry boulders from the stock, but also some rounded boulders of granite and gneiss. The porphyry came from Hahns Peak but the rounded boulders of granite and gneiss were derived from one of the conglomerates.

Granite and gneiss occur at Farwell Mountain, six miles east of Hahns Peak village, and in the Park Range, six miles farther east. At both places quartz veins carrying a little gold have been found. Presumably these or similar sources provided the boulders found in the conglomerate. Most of the placer gold is fine one millimeter or less—but some pieces of coarser gold and some gold attached to quartz have been recovered. No gold-bearing quartz veins have been found in Hahns Peak.

With this limited, but conflicting, evidence it is not surprising that arguments have arisen about the source of the gold—whether Hahns Peak porphyry and marginal altered rocks, the older conglomerates or the veins in the mountains to the east are the immediate source. I believe that the porphyry and its marginal altered rocks or perhaps deposits related to them, undiscovered or even now eroded away, are the immediate and the original sources of the placer gold. The deposits are in streams which radiate from the Hahns Peak intrusive. If the conglomerates were the source, since they are widespread there should be similar placers elsewhere and there are none. If the quartz veins of the Park Range were the immediate source, it is difficult to explain why there are no placers between Hahns Peak and that source.

Hahns Peak has not been actively glaciated, although it probably had hanging ice fields. There have been many landslides in its gulches and debris from them has accumulated at the foot of the peak. Placers have been traced and worked into these accumulations, but no placer deposits have been found above them.

The gold found in the placers is mostly small—less than one millimeter in size. It is light yellow but contains considerable silver; fineness appears to be in the 600–750 range.

Select your samples at Poverty Bar or in Ways Gulch or on Willow Creek. Poverty Bar is a remnant of an alluvial fan. Its gold will be concentrated in local channels at various levels; each channel should be marked by a lens of coarser fragments in the gravels. The other two are stream placers in which gold is concentrated near bedrock. On Willow Creek there may be skim bars. Reportedly, drilling on Willow Creek and lower Ways Gulch showed good values, but the bedrock gravels are probably too deep for you to reach. The placers in Little Red Park are colluvial deposits with chaotic gold distribution. They are difficult to find and to reach and aren't recommended for a first try.

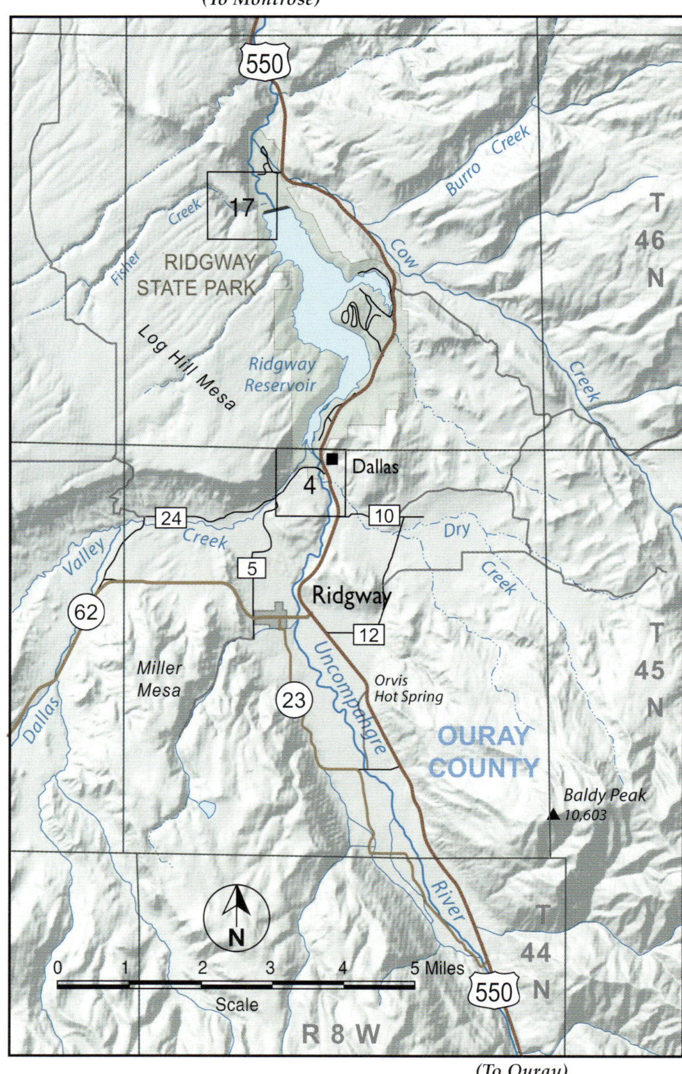

Figure 53. Location map of Ouray County placers.

SAN JUAN COUNTY

All the major valleys of San Juan County were glaciated in Wisconsin time and the placers they may have contained before were destroyed (see page 30). The little placer gold they contain has been deposited in post-glacial, and even in modern, time. Although the county has been a major gold producer, placer mining has never been important here. Nevertheless the area can be of interest to the panner.

One of the first gold discoveries of the San Juan Mountains was made by Charles Baker's group of six men in 1860, near the site of the old town of Eureka in NE sec. 19, T. 42 N., R. 6 W., (Unsurveyed). The following year a group of over 100 men arrived and placered in Eureka Gulch and at its mouth on the Animas River. They could not make wages and the placers were abandoned by the end of the season.

The valley gravels do contain some gold and many of the important bedrock deposits were found by prospectors following faint trails of colors upstream and then upslope to the outcrops. The story of a modern discovery—the Corey deposits high on the slopes above South Mineral Creek—is told on page 1.

There were many stamp mills in the county when the mines were being worked. Some of them were large and some worked for many years. Many of them were very inefficient and all of them lost more or less gold in their tailings. A part of the gold found in today's gravels is stamp mill gold, and in at least one locality a small but rich "man made" placer was generated. This was on Cement Creek immediately below the Gold King mill in sec. 21, T. 42 N., R. 7 W.

The gold in the modern placers, as might be expected from its having been liberated by glacial processes or by stamp milling, is very fine-grained. Part of it probably accumulates on skim bars at times of low water, and much is swept away.

Select your samples from near-surface gravels on the major streams and from skim bars on the Animas River. Don't neglect trying tributary gulches, especially Eureka and Arrastra Gulches. Below Silverton the gravels of the Animas River are gold-bearing and some of the lefthand tributaries (on the east side) downstream are worth prospecting. It's worthwhile to inquire about the old mills and check the slopes below them. The mills should have operated before World War I—most later mills lost less gold.

SAN MIGUEL AND MONTROSE COUNTIES

Individual placer operations years ago are reported on Big Bear and Bilk Creeks in T. 42 and 43 N., R. 10 W., and placer gold is said to have been found on the upper part of Mesa Creek drainage, probably in the southern part of T. 49 N., R. 17 W. The workings must be small, for I was never able to locate them. The placer production of San Miguel and Montrose Counties has not been large, and almost all has come from the Keystone and San Miguel-Dolores River placers. Of these two, the Keystone placer has probably accounted for the greater part of the gold.

Some of the State's largest gold mines are located in the upper parts of the San Miguel and South Fork drainages. However, as is the case in Ouray and San Juan Counties, the gold placer deposits developed from them are not nearly so extensive as those derived from some much less productive districts. This is true of all the gold districts of the San Juan Mountains (see page 30).

The first significant placer found going downstream from the head of the San Miguel River is the Keystone placer, at its junction with the South Fork, nearly four miles downstream from Telluride and more than five miles downstream from the nearest important bedrock gold deposits. Only a few small, very low-grade placers that did not pay to work were found by early prospectors on the flat east of town between the Keystone placer and Telluride. From place to place in the streams above Telluride and on the slopes they must have found colors leading to some of the vein outcrops, but they found no placer deposits worth working.

Probably there is now some modern stamp mill gold on skim bars, but significant modern placers should not be expected immediately below the old mills as occurred at the Gold King mine in San Juan County. This is because the discharge streams of most of the mills here were so overloaded that placer concentration usually did not occur and also because, when cyanide extraction of gold was developed, plants were built here that treated most of the accessible stamp mill tailings accumulations.

Keystone Placer

The Keystone placer is in the NE sec. 32 and NW sec. 33, T. 43 N., R. 9 W. The first work here was in 1881 and some drifting and small-scale hydraulicking continued each season until 1892. Afterward there was little activity until 1901 when construction of dams, flumes and pipelines for the present large cut's operation began. Mining began in the fall that year and continued until 1906, when large landslides into the placer cut stopped operations.

The cut walls that can be seen are in morainal material from the San Miguel and South Fork valleys. (Glaciers in the two valleys coalesced here.) The highest point on the faces of the cut is 430 feet above the river bed beside the cut. The morainal material contains traces of fine gold, but most of the gold produced here was coarse, concentrated on and just above bedrock where enormous boulders gave the miners many problems. It appears that at Keystone the glaciers overrode extremely coarse, rich alluvial debris. Because of landslides, this bouldery debris is no longer exposed, although on the floor of the pit there are some giant boulders left by the miners.

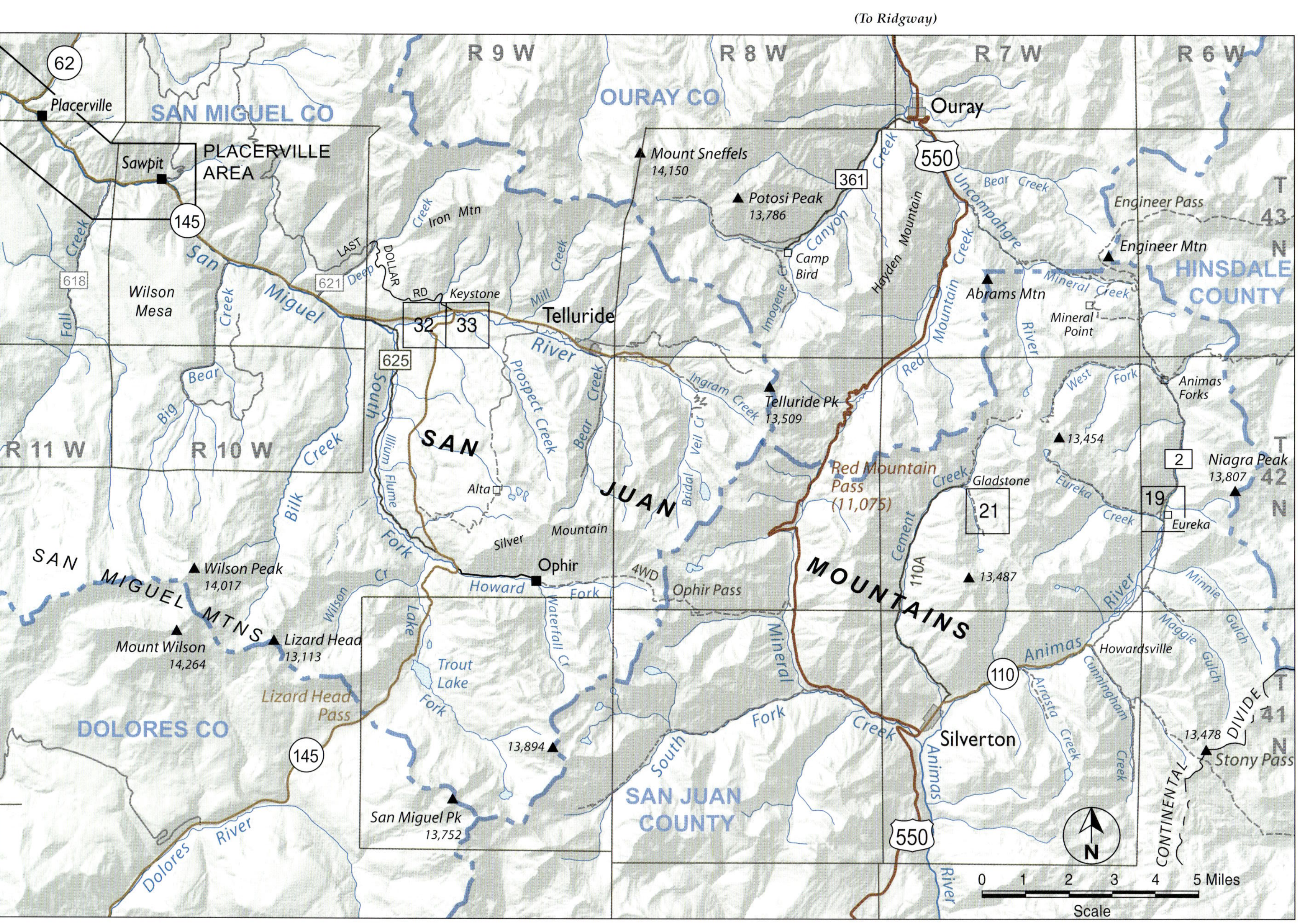

Figure 54. Location map of San Miguel and San Juan County placers.

Here select for panning stream bed gravels just below the hydraulic cut which may be richer than the moraine material in the cut walls, A second choice is any coarse, cobbley lenses you may find in the moraine which might be meltwater channel gravel.

San Miguel-Dolores River

The San Miguel River heads in Bridal Veil and Ingram Basins southeast of Telluride, flows through Telluride, Keystone, and Placerville in San Miguel County, flows through Naturita and Uravan in Montrose County and joins the Dolores River approximately three miles downstream from Uravan. In Montrose County it is joined by Cottonwood Creek in sec. 3, T. 46 N., R. 14 W., and by Roc Creek in sec. 4, T. 48 N., R. 18 W.

Most of the placer work has been done in three areas: on the San Miguel downstream for six miles from Sawpit, downstream from Cottonwood Creek for about six miles and on the Dolores for nine miles upstream from Roc Creek. Placerville is near the center of the first group of placers; Naturita—several miles below the second group; and Uravan—just upstream from the third group.

The placers in the Placerville area probably were discovered in 1875. They were worked from 1876 through 1882 and occasionally thereafter. Only a little work was done during the 1930s. Probably this was because of the difficulty in getting water to the terraces or the gravels to water, for the gravels here are apparently higher grade than those of the placers downstream.

In 1883 several companies were formed to work placers in the two downstream areas. One of these was the Montrose Placer Mining Co., with properties at Mesa Creek in the lowest area, which built the spectacular "hanging flume" in the Dolores River Canyon. (Remnants of the flume can still be seen in the canyon. See page 20.) The company's water supply was provided by a ditch and flume over 10 miles long, about half ditch and half flume. It traversed the Dolores Canyon, a distance of about four miles, in a flume six feet wide and four feet deep pinned to the sheer rock of the canyon wall. In one stretch several hundred feet long where the wall overhangs, the flume hung from pins set in the rock overhead. The flume in the canyon was 100 to 150 feet above the river and several hundred feet below the edge of the canyon.

A road was built along the lip of the canyon and workmen hung from the top on ropes to mark the level of the flume on the wall. The holes were drilled and the flume constructed from a derrick on the flume floor that was advanced as flume sections were completed. A construction crew of 12 men was able to build up to 250 feet of flume per day, The ditch and flume were built upstream from the placer site, since it was nearest the lumber source.

The property was tested in the spring of 1888, the upper end of the ditch and flume constructed in 1890, construction of the tail sluice below below the placer began early summer of 1891, the first cleanup was made about September, 1891, and the Company suspended operations in 1893. At the time it was reported that the flume cost "something over $100,000", consumed 1.8 million board feet of lumber and was built in "more than two years."

The Montrose Company was one of five companies that had projects at the Mesa and Cottonwood Creek areas and one of two that actually operated. It brought water to the Mesa Creek area; the U. S. Gold Placers Company brought water to high bars opposite Cottonwood Creek in a nine-mile ditch. Both companies operated about a year and failed. In each case the testing, if any, had been poorly done and the value of the placer gravels greatly exaggerated.

There was no further placering until the 1930s when there were a number of small-scale operations on the rivers.

The river canyons are cut in sedimentary rocks with alternating hard and soft sections. In the three placer areas there are elevated terraces beside the river, each floored on hard sandstone beds at most places. There are many remnants of the terraces on the slopes of the canyons.

The prominent terraces are:

	Heights above river in feet		
	Placerville area	Cottonwood Creek area	Mesa Creek area
Terrace	180		
Bedrock	160		
Terrace	385	120	
Bedrock	215	35	
Terrace	210–270	190	60–85
Bedrock	110	140	40–45
Terrace	150–185	27–45	25–30
Bedrock	80–85	12–20	

These terraces have been worked in cuts at many places, and in many of them the gravels can be seen or reached without too much cleaning. The terrace gravels have been worked in drift mines in many places. These drift mines are very old. They are unsupported, since they did not need timbering when they were driven, years ago. ***If you find any open, stay out of them. They are very dangerous!***

Figure 55. Part of the Keystone Hydraulic Mine, San Miguel County, the second largest hydraulic mine in the State. The pit wall can be seen on the right. Beyond it is the flume supplying water-650 feet above the pit floor. Near the house the water flow was divided, with several lines to the right of the house feeding monitors (one is at the bottom right) and one line to the left of the cabin going into the sluice. The San Miguel River and the sluice are in a depression we cannot see between the cabin and monitor jet. In normal operation the monitors played on the walls of the pit to the right of the photo. (Denver Public Library, Western History Department)

The gold in the river gravels is fine-grained. It is somewhat less so in the terrace gravels and less fine-grained at Placerville than downstream. The gold in the terrace gravels is concentrated near bedrock; toward the terrace surfaces the gravels are nearly barren. The gold in the stream gravels is concentrated in the upper few feet of gravel at skim bars. Bedrock beneath the modern river floors has never been tested.

Select gravel samples from skim bars along the modern rivers in the three areas described above. You can also get samples for panning from as close to bedrock as possible in cuts on the terraces. The table gives an idea how high to climb. There is no water on the terraces, so you must pack either gravel down or water up. I suggest the first. ***Once again, don't enter any drift workings!***

SUMMIT COUNTY

Bemrose Placer Area

Several placers have been worked in NE sec. 1 and central sec. 12, T. 8 S., R. 78 W., at the head of the Blue River near the foot of Hoosier Pass. In many places the gravels are covered by several feet of peat, which the miners had to strip. They are unusual in that the placers in sec. 12 are in moraine from the South Platte glacier that spilled over the top of Hoosier Pass. (On the south side of the pass the South Platte valley abruptly turns 90 degrees, flowing eastward from the high peaks of the Tenmile Range and then southward below the pass.) The placers in sec. 1 are in lateral moraine of the Monte Cristo Glacier that contains material from the South Platte Glacier.

McNulty Gulch

A small but apparently very rich placer, discovered in 1860, was worked in McNulty Gulch in the SSE sec. 35, T. 7 S., R. 79 W., and NE sec. 2, T. 8 S., R. 79 W. from beneath the Climax tailings pond upstream almost three quarters of a mile It is reported that bedrock gravels yielded as much as five ounces to the pan and that a 22-ounce nugget was found here in 1867, about at the edge of the pond. Gold occurs in thin gravels on bedrock and two to three inches into it. The gravel lies beneath glacial till in the lower part of the placer and was worked there by drift mining. The gold is coarse and not flattened. Its source has never been determined.

Follette Placer

The Follette placer is in SW sec. 13, T. 7 S., R. 79 W., It lies to the southwest of Colorado 91 and northeast of Mayflower Creek. It is on an abandoned channel of Mayflower Creek in the glacial outwash terrace 75 feet above Tenmile Creek. There are some small placer pits between the Follette placer and the mouth of Clinton Creek. There were some beneath the reservoir in Clinton Gulch. Although none is rich, the gravel of the Follette placer is probably better than any in Clinton Gulch,

Mosquito Range

Gold-bearing mantle—fine material containing coarse gold in the crevasses between talus blocks—has been found on the west slope of the Tenmile Range near its crest in sec. 33, T. 6 S., R. 78 W.

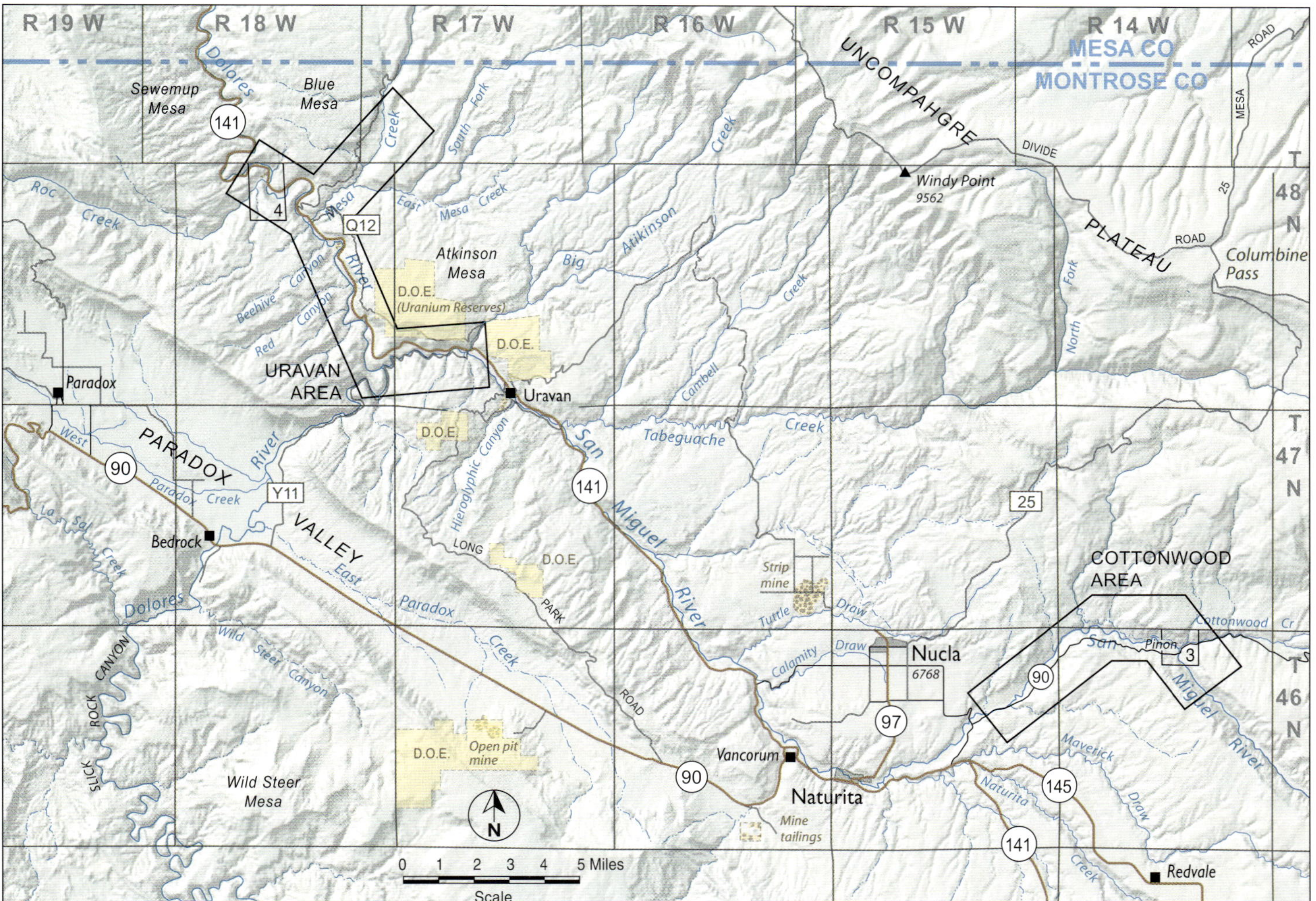

Figure 56. Location map of San Miguel-Dolores River placers, Montrose County.

TELLER COUNTY

Placers have been found at a number of places in the Cripple Creek district, but their production has been small—probably aggregating no more than 1,200 ounces. Reportedly some of the placers discovered were left unworked because of their small size and the general lack of water. There were small operations in 1894-1897 and probably for a few years thereafter. Again, throughout the 1930s there were small operations and in 1951 there was a small dry land washing plant in operation.

The gold recovered was angular, mostly bright yellow but with some fragments with a rusty surface, and it contained very little silver.

The largest placer area was on the southwest slopes of Mineral Hill and in Pony Gulch north and west of Cripple Creek town. Because the gravels are thin and the deposits spotty, the small placer dumps are not very noticeable. The gold clearly originated in Mineral Hill, but the bedrock sources have never been recognized.

Some placers were worked south of the town on Cripple Creek, and in Squaw and Arequa Gulches. The name "Gold Run" suggests that some placers were found there.

Placers were worked at Hull's Camp, near the center of SE sec. 20, T. 15 S., R. 69 W., and on Wilson Creek south of Victor. There was an eluvial placer on Globe Hill in SE sec. 18, but its gravels were treated by milling.

Collect dirt for panning on the slopes of Mineral Hill and the head of Pony Gulch. This is the easiest area to locate and had the most extensive placers. From the foot of Mineral Hill up its slopes, the gold is in random distribution. In the broad valley of Pony Gulch and downstream it is concentrated on bedrock. Many of the other sites were once built up, and it will require some patience to find undisturbed gravel.

SELECTED REFERENCES

Conrow, Robert, 1983, The great diamond hoax and other true tales: Boulder, Colo., Johnson Books, 62 p.

Ellis, E. H., 1967, The gold dredging boats around Breckenridge, Colorado: Boulder, Colo., Johnson Publishing Co., 173 p.

Henderson, C. W., 1936, Mining in Colorado, a history of discovery, development and production: U. S. Geological Survey Professional Paper 138, 263 p.

Hollister, O. J., 1867, The mines of Colorado: Springfield, Mass., Samuel Bowles and Co., 450 p.

Parker, Ben H., Jr., 1961, The geology of the gold placers of Colorado: Colorado School of Mines Doctoral Thesis, 578 p.

______1974, Gold placers of Colorado: Colorado School of Mines Quarterly, v. 69, nos. 3 and 4, p. 268 and 224.

Peele, Robert, ed., 1941, Mining engineers' handbook, 3d edition, v. 1, John Wiley & Sons, 541 p.

Singewald, Q. D., 1950, Gold placers and their geologic environment in northwestern Park County, Colorado: U. S. Geological Survey Bulletin 955-D, p. 103–172.

Wells, John H., 1969, reprinted with errata 1989, Placer examination, principles and practice: U. S. Bureau of Land Management Technical Bulletin 4, 209 p.

Woodard, Bruce A., 1967, Diamonds in the salt: Boulder, Colo., Pruett Press, 200 p.

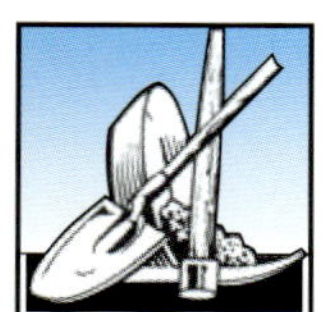

MAP INDEX BY COUNTY

Pages on which you can find location maps of a county or part of a county